# TECHNOLOGY FOR TOMORROW

**Charles F. Williams**
Associate Professor
Science Education and Teacher Education
California State University at Fullerton
Fullerton, California

**Kamiran S. Badrkhan**
Assistant Dean
Electronics/Computer Science
Los Angeles Trade-Technical College
Los Angeles, California

**Willard R. Daggett**
Director
Division of Occupational Education Programs
New York State Education Department
Albany, New York

P25

Published by
SOUTH-WESTERN PUBLISHING CO.

CINCINNATI WEST CHICAGO, IL DALLAS PELHAM MANOR, NY PALO ALTO, CA

ISBN: 0-538-16250-3

Library of Congress Catalog Card Number: 84-50129

1 2 3 4 5 6 7 8 9 K 9 8 7 6 5 4

Printed in the United States of America

# CONTENTS

**Preface** .................................... **vii**

## PART I. INTRODUCTION

**1. The Effects of Technology** .................... **4**
People and Needs 5
Problems and Concerns 9

**2. Technology and Change** ...................... **24**
Civilization and Technology 25
Production 27
Transportation 33
Health 38
Structures 39
Communication 42

**3. Human and Natural Resources** ................ **48**
The Vital Importance of Resources 49
Human Resources and Technology 50
Natural Resources and Technology 57
Systems of Technology 66

## PART II. PHYSICAL TECHNOLOGY

**4. Tools and Mechanisms** ...................... **72**
People and Tools 73
Tools 76
Mechanisms 83

**5. Power** .................................... **92**
The Nature of Power 93
Systems of Power 95
External Combustion Systems 96
Internal Combustion Systems 99
Fuel Usage 104
Related Technologies 105
Different Types of Power 107
Relationship of Power to Energy 109

**6. Energy** .................................... **112**
The Nature of Energy 113
States of Energy 115
Harnessing Energy 118
Nature of Electricity 119
Electrical Batteries 121
Generating Electricity 122
The Power-Energy Connection 126

**7. Structures** .................................... **130**
Meeting the Need for Shelter 131
Structural Forms 134
Framing 143
Exteriors 144
High-Rise Buildings 144
Perimeter-Supported Buildings 146
Bridges 147
Shell Structures for Vehicles 148
Form and Function in Structures 151

**PART III. BIOTECHNOLOGY**

**8. Agritechnology: Products** .................... **156**
Scope of Agritechnology 157
Agritechnology Methods 160
Agricultural Commodities (Crops) 161

**9. Agritechnology: Methods** .................... **182**
The Technologies of Agriculture 183
Land Preparation 183
Planting 188
Cultivation 188
Irrigation 190
Harvesting 193
Storage and Handling 196
Processing 198
Breeding and Development 201

**10. Health Care** .............................. **206**
Health and Its Care 207
Technology and Life 208
Life (Vital) Signs 214
Life Cycle 217
Threats to Health 218
Technologies of Health Care 226
Diagnostic Technology 227
Treatment 230

**11. Health Maintenance** .............................. **234**
Technologies for Health Maintenance 235
Personal Health Maintenance 238
Exercise 243
Physical Therapy 247

## PART IV. INFORMATION TECHNOLOGY

**12. Information Systems** .............................. **254**
The Importance of Information 255
Sources of Information 256
Computers, Data, and Information 258
The Information Explosion 263
The Information Processing Cycle 268

**13. Computer Systems** .............................. **276**
Computers and Information Processing Systems 277
Computer Operations 286

**14. Computers and Change** .............................. **304**
Technology and New Ideas 305
The Effects of Computers 313
Personal Services 317
Office Automation 320
Robotics 324
Ergonomics 326
Artificial Intelligence 328

## PART V. PUTTING TECHNOLOGY TO WORK

**15. The Systems Approach: Putting Technology to Work** .............................. **334**
Development of Technologies 335
Systems of Technology: Teamwork in Using Knowledge 336
The Systems Approach 338
Developing Technological Systems 346

Set Goals 348
Evaluate Feasibility 351
Establish a Plan of Work 354
General Design 357
Technical Design 359
Development 360
Implementation 361
Evaluation 362
Management and Feedback 363

**16. Technology at Work** .......................... **366**
Technology and Opportunities 367
Agriculture 368
Infrastructure 370
Production 371
Materials Handling and Transportation 373
Aerospace/Military 376
Communication—Transmission 378
Communication—Information 378
Commerce and Banking 380
Education 381
Administration 382
Personal Services/Life-Style 383

**Glossary** .......................... **386**

**Index** .......................... **399**

# PREFACE

*Technology for Tomorrow* establishes a new outlook for occupational and career-related education. The purpose of this book is to help students acquire a background and perspective on the technologies within which they eventually will live and work. Therefore, this basis of knowledge should help shape the directions of future educational and career choices made by students.

Traditionally, the diverse areas of occupational education have taken what can be characterized as a smorgasbord approach. That is, students were led through arrays of information and experiences dealing with a variety of occupations. It was up to the student, with some help from teachers and counselors, to pick and choose.

The new approach, implemented through this and companion works, centers upon students, rather than jobs or working techniques. As a starting point, this book establishes a broad base of knowledge about what technologies are and how technologies are combined into systems that serve people. Students acquire knowledge of what has happened in the recent technological past. They also build an understanding and appreciation for the fact that change has become a way of life in modern society. As a next step in this new approach, students will be able to go on, in a subsequent text, to a review of direct career and job implications of systems of technology.

## CONTENT ORGANIZATION

*Technology for Tomorrow* is organized into five parts and 16 chapters.

### Introduction

This first part consists of three chapters that establish an awareness of technology. The relationship of technology to change and the dependence of technology upon natural resources are stressed.

Chapter 1 begins with a discussion of the basic nature of technology. Technologies evolve to meet human needs or wants. Throughout history, people have always required or aspired to more than they had at a given point in time. This concept is critical for student understanding of what technologies are and why technologies are developed. Some people have recognized human needs or wants as opportunities. These people have adapted natural resources and developed techniques for providing the essentials for life and the additional comforts that have been in demand. The chapter then introduces the major problems and concerns of people that have been addressed by technologies: safety, life support, health, comfort, and knowledge. These factors are covered as previews of the technological topics to follow.

Chapter 2 carries the title *Technology and Change* and is historically oriented. The text describes the pattern of evolving technologies as people moved from caves and learned to use crude tools to urban living and use of power devices. The main thrust of the chapter is that changes occur because people are ready for new methods or have established demands. Types of changes covered in this chapter deal with the areas of production, transportation, health care, structures (buildings), and communication.

Chapter 3 covers the relationships of human and natural resources to the evolution of technologies. Topical breakdowns used as a basis for these discussions include demand, invention, implementation, and control. The point is made that, at the outset, technologies meet human demands. It also is pointed out that technologies can have adverse side effects. People devise controls to deal with these. The chapter discusses the roles of natural resources, including natural sources of energy and critical materials, as they relate to technology.

## Physical Technology

The four chapters in this part of the book deal with areas of technology that are related most closely to traditional industrial arts education. This organizational plan has been used with a purpose: Established industrial arts teachers already have the experience and the equipment to demonstrate and deal with most of the content involved. The chapters in this part cover tools and mechanisms, power, energy, and structures. For students taking parallel work in sciences, these topics provide a practical, applied dimension for lessons derived from courses in physical science.

Chapter 4 deals with tools and mechanisms for doing work. Tools are identified as objects that are used directly by people to perform specific tasks delivering identified results. Examples include hammers, wrenches, saws, and other cutting devices. Mechanics, by contrast, involves the principles and devices by which people multiply the work they can do. Mechanisms described in the chapter include the lever, the wheel, the inclined plane, the pulley, and gears.

Chapter 5 deals with power. The chapter discusses the technology of power. The primary instruments of power that are described include external and internal combustion engines. Students are led through a description of the operation of a typical automotive, four-stroke cycle engine. Discussions also cover the diesel engine, as well as rocket and turbine engines. Students are made aware of the relationships between the generation of power and the consumption of important natural resources—fossil fuels.

Chapter 6 discusses energy. The chapter begins with a simple, everyday example. The act of sunbathing is related to the use of the most prevalent source of energy, the sun. The chapter then discusses the states of energy: kinetic, potential, and rest. Within this context, energy centers around the generation and use of electricity. The generation of electricity through use of electromechanical devices then is discussed. Students are given a basic grounding in magnetism and electricity. The generation of electricity is then related to the operation of mechanical devices described in the previous chapter.

Chapter 7 deals with structures, running all the way from early cave dwellings through modern meeting places, factories, and homes. The principal portions of structures, including foundations, frames, and exteriors, are covered. Types of structures, such as high-rise buildings, perimeter-supported buildings, bridges, and shell structures are described.

## Biotechnology

The third part, consisting of four chapters, takes the students through an important transition in their review of technologies. This part deals with living things, including human life. Students are helped to understand that such areas as human health and agriculture have been principal focuses of technology through the years. As growing emphasis is placed upon the control of illness and upon expanding agricultural production, the related technologies are expanding. These areas, therefore, represent potential career opportunities with which students should be familiar.

Chapter 8 covers one portion of a broad topic identified as agritechnology. The placement and content of this chapter are significant to the book and to the course it supports. Studying agriculture represents a natural entry point for students at the age of the projected readers for this book. Teen-age students may be inclined to shyness in studying the mechanisms of the human body. Indoctrination through agricultural life removes some inhibitions.

As another conceptual framework, this chapter isolates the study of the products of agriculture and their nature from concerns about the methods used in farming and food processing.

The chapter identifies agricultural crops (commodities) and discusses their nature, structure, and potential importance to people. Students learn that there are some 3,000 plants that can be cultivated as crops. However, the great majority of the food consumed by people falls within 16 species. These species are then reviewed as the main topics for this chapter.

Chapter 9, building upon the content of the previous chapter, deals with the technologies of agriculture. Topics include land preparation, planting, cultivation, irrigation, harvesting, storage and handling of foodstuffs, processing, and breeding and development of both plants and animals.

Chapter 10 deals with health care. Presentations are made within a technological framework. That is, the human body is described in terms of a number of its vital systems, including circulatory, respiratory, nervous, muscular-skeletal, digestive, excretory, reproductive, and endocrine. The monitoring of the health of the human body is described in terms of vital signs which are identified and described. These include pulse, blood pressure, respiratory rate, and temperature. Health care requirements then are related to the human life cycle. It is explained that requirements for health care are greatest for infants and the elderly, with health care requirements diminishing for youths and young adults. Technologies for diagnosis of human health are described.

Chapter 11 introduces the idea of health maintenance. Whereas health care deals with sickness, health maintenance includes measures to avoid illness by maintaining the body in a healthful state. Options and methods covered include health maintenance organizations, medical checkups, preventive medicine, and holistic medicine. Sections of this chapter also deal with nutrition and with the value of exercise.

## Information Technology

The content within this fourth part of the book can be used to meet introductory-level requirements for computer literacy that have been introduced in some states and local jurisdictions. Overall, the three chapters in this part deal with information as a human resource and with methods for providing required information. The methods include both the handling of information and the use of computers. Information systems concepts are introduced as a prelude to further, in-depth discussions about systems management in the final part.

Chapter 12 is about information systems. It is explained that data, or raw, basic facts, are processed to produce information. Information, in turn, is used to generate knowledge. Within this context, computers are presented as tools for the processing of data and the development of information. Students are led through a historic review of the evolution of data processing equipment, leading up to the introduction of computers. Then, the basic elements of information systems are presented: input, processing, output, and storage. This information processing cycle is described as essential to all methodologies for handling data and creating information.

Chapter 13 is about computer systems. Systems of technology are shown to consist of a number of necessary functional parts, including people, data, procedures, and processing equipment. The role of people is described as paramount. People design and use computers and computer systems. Data are shown as resources of an information society. Procedures are the things that make systems work. Processing equipment, including computers, are technologically evolved tools.

Chapter 14 deals with the phenomenon of computers and change. Throughout the book, it has been pointed out that technologies meet human needs or wants. Once a technology is introduced, it becomes an agent of change on its own. That is, the new products of technology become the focal points of methodologies capitalizing on additional opportunities. At this point, the computer is used to illustrate this pattern.

This chapter also deals with the issues of both employment opportunities and potential displacement of jobs that have resulted from the rapid introduction and acceptance of computers.

### Putting Technology to Work

This fifth part of the book has two chapters dealing with the development of technologies and with the effects of technologies upon employment opportunities.

Chapter 15 is about the systems approach and systems management in the development of technologies. It is pointed out that technologies result from major projects involving efforts of many people and the use of considerable financial and natural resources. To develop a new technology or to combine capabilities of existing technologies, a systems approach is typically used. This involves a project structure that is identified and described. Steps in this systems process are given: goal setting, feasibility evaluation, establishing a plan of work, general design, technical design, development, implementation, and evaluation.

Chapter 16 deals with technology and the workplace. A number of career areas impacted by technology are identified. Discussions of these areas describe both opportunities and limitations for future careers. The areas covered include agriculture, infrastructure, production, materials handling and transportation, aerospace/military, communication-transmission, communication-information, commerce and banking, education, and administration.

## THE PLACE OF TECHNOLOGY STUDIES

The concluding chapter of this book, in effect, prepares students for learning experiences that should follow. That is, a companion text, entitled *Technology at Work,* concentrates upon the identified job-related areas described in the final chapter of this book. Thus, the book and the course of which it is a part should establish a continuity of learning for students. The base of knowledge students have formed can be used next for a forward-looking review of the opportunities that will be open to them with further study and experience.

The study in this program should be related to a course sequence that involves the principles of citizenship, personal development, and career selection. The study of technology should supplement and enhance these vital, basic curriculum offerings.

## ACKNOWLEDGMENTS

Significant assistance in the development of the manuscript for this book was realized through contributions made by an outstanding team of reviewers. The help rendered by these individuals is hereby recognized. They are:

Richard Kitzmann, Supervisor, Industrial Education, State Department of Public Instruction, Madison, WI

Fred Posthuma, Westfield High School, Westfield, WI

Michael Hacker, Syosset Junior High School, Syosset, NY

Leslie Alexander, Bloomington High School, Bloomington, IL

# 1 INTRODUCTION

You live at a time that has been called an age of technology. Technology is part of your everyday life. This book is aimed at helping you to understand and benefit from the technologies in your life.

Technology is a method of putting the principles of science to work to help people. Science, in turn, is the study of life, earth, and physical things. The lessons of science are the building blocks of technology.

The first chapter in this part of the book deals with the relationships of people and technologies. Technologies meet needs or wants of people. If the wants and needs of people didn't come first, technologies could not succeed. In particular, technologies have been in high demand as people have crowded into cities. Life in cities, urbanization, requires technologies. Without technologies, there would be no large towns or cities.

The second chapter relates technology to change. Technologies, by themselves, do not cause change. However, once a technology exists, it can become an agent for promoting change. Living with change, in turn, is one of the major challenges of modern life.

The sources of technology are described in the third chapter. Technologies use and rely upon natural resources. Technologies also help people to deal with shortages in natural resources.

# 1 THE EFFECTS OF TECHNOLOGY

### Your Learning Job

After you have read this chapter and completed the work assignments, you should be able to:

- ☐ Define technology.
- ☐ Explain how progress is directed by human needs and wants.
- ☐ Explain why technology is a tool to help people, rather than a force for change.
- ☐ List and describe five major concerns of people to which technology is applied.
- ☐ Explain how knowledge is cumulative.
- ☐ Define inductive and deductive reasoning and explain their differences.
- ☐ Describe the delphi process.

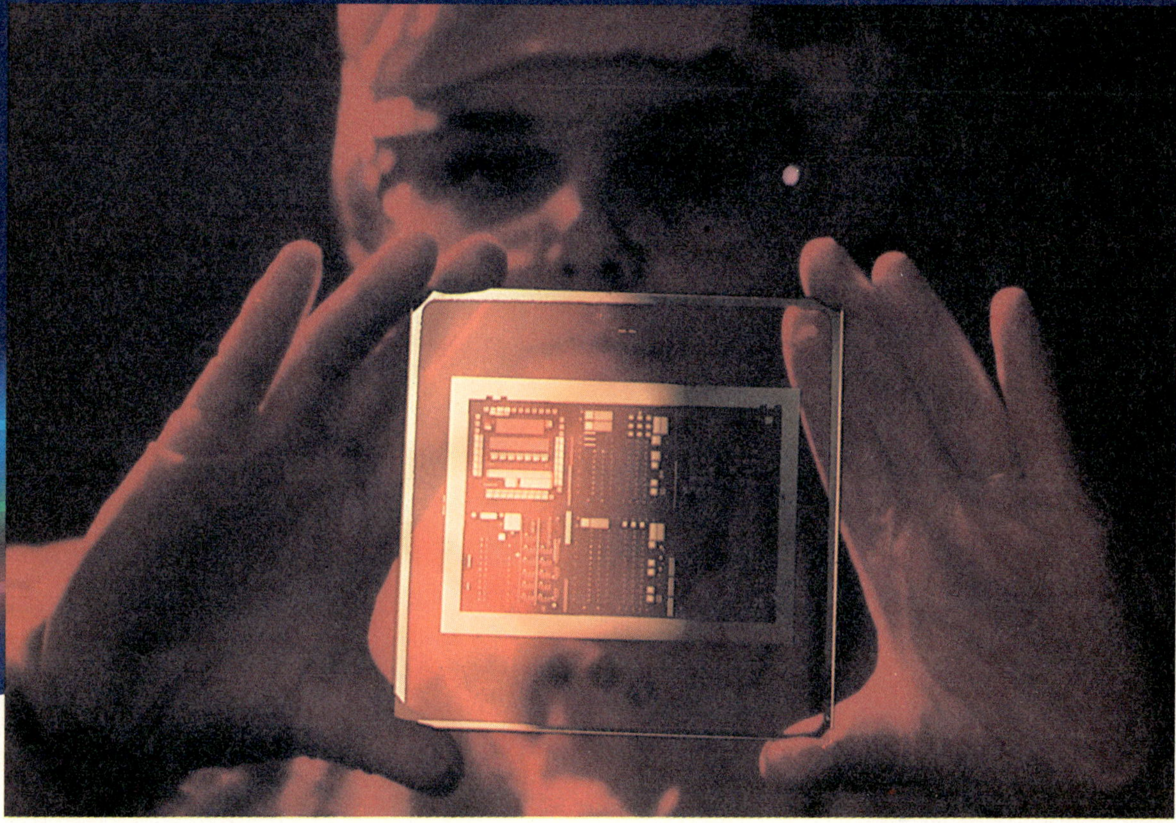

The future takes shape through systems of technology. This photo shows a film master from which computer chips will be made. Computer chips represent a combination of photographic and electronic technologies, as well as some special designs for the handling of data and information. (Courtesy Sperry Corporation)

## PEOPLE AND NEEDS

People dream. The dreams people dream depend upon where they are and the comforts they enjoy. In some parts of the world, people's fondest dreams may deal with having enough to eat, clothes to wear, or places to live. In other parts of the world, possibly for many people in the United States, dreams may be out of this world. Undoubtedly, many Americans are already dreaming about traveling to the moon or setting up colonies in space. Other people in the United States may be looking forward to traveling to exotic resorts or to buying new cars.

Thus, the dreams people dream depend upon some very human factors. Hungry people dream about food. People who are cold dream about houses and clothing. People who have taken care of their basic needs for life support may dream about luxuries and adventure.

People have always dreamed of a better world—since the beginning of time. The definition of a better world changes as people's situations change. Over time, people have invented things and made things that have been lumped together under the name of *progress*. Many of the comforts you count on in your everyday life result from some form of progress. The paper in your books or notebooks represents a basic kind of progress on which knowledge is recorded. Progress has also given you and your family such things as cars, radios, television sets, telephones, and airplanes. These are the comforts for today.

**The needs and wants of different people** depend on where they live and what they have. The less people have, the greater their needs will be. Whatever their needs and wants, people depend upon technology to help meet them. (UNRWA)

Now, think back. Imagine, dream about, a young person in prehistoric times. The family lived in a cave. In these times, a person who was 14, 15, or 16 would be considered middle-aged. That is, people lived to perhaps 30. The world was very basic. Comforts were virtually unknown.

People dreamed about warmth, about being able to find a dry place to live, about having food to eat and clothing to wear.

The imagination and the actions that went along with these early dreams were simple and practical. Picture what might have happened: Lightning strikes suddenly. A nearby tree is knocked down and begins to burn. Fascinated, people move closer and observe the fire. As they do, they feel warm and comfortable. Somebody thinks: Why not take a stick from the burning tree into the cave? Then, other sticks could be added, and a fire could be kept going to warm the living area.

Progress can be as simple and basic as learning to use fire to warm a dwelling. Progress can be as complex and advanced as developing the computer. Both innovations share one thing: The effects on people have been dramatic and far-reaching. Progress means change, and when change occurs, the way people look at the world, their outlook, changes. For the cave dweller, the idea of using fire for heat improved the quality of life. Such other ideas as floating on water for transportation or carving wheels out of logs made work easier. In later times, human progress was measured by such developments as motion pictures, television, telephones, or airplanes.

Clearly, people at different times through history were ready for different levels of progress. The progress that could be made at any time was based on what people were ready to understand and to do. Putting it another way, progress depends on how people live, what they know, or their outlook. All of these factors—knowledge, human activities, and needs or wants that people identify—fall into a single category. Today, the knowledge and activities that achieve progress are known as *technology*.

**Putting a man on the moon** was an old dream that came true in the 1960s with the help of space technology. (Courtesy NASA)

People have always had needs and wants. They always will. The ability to meet needs and wants depends upon what people have and what they know. The more people have—and the more they know—the more they will always need and want. Human nature is creative and will always strive to make dreams come true.

## Technology and Dreams: The Urge to Fly

Take flying. People have dreamed about flying for many centuries. A Greek myth tells of a father and son, Daedalus and Icarus, who wanted to fly. They fashioned wings from wax.

When the wings were built, Icarus found that he could fly like a bird. Icarus was young and eager—and not anxious to heed advice from elders. Daedalus, a wise father, warned him not to fly too close to the sun. But Icarus was anxious to soar higher and higher. Eventually, Icarus flew so high that the heat from the sun melted the wax wings. The wings were destroyed. So was Icarus. But the dream lived on.

**Dreams of flight** go back to ancient history. Greek mythology tells the story of Icarus (below) who burned his wax wings by flying too close to the sun. Above, Daedalus, his father, watches anxiously. (Courtesy National Air and Space Museum, Smithsonian Institution)

The great artist/scientist/innovator, Leonardo da Vinci, sketched flying machines as long ago as the fifteenth century. Leonardo's journals contain drawings of winged vehicles displaying some of the same principles ultimately built into twentieth-century aircraft. Centuries later, people finally did fly. Eventually, airplanes became the main vehicles for moving people long distances.

**The jet airplane,** a product of a system of technologies, is the main method of long-distance travel. (Photograph provided by Delta Air Lines)

The dream of flying presents the same lesson as other areas of technology: People have to need or want something first. Technology is a *solution,* an answer, not a force for change. However, technology does lead to change. Further, the greater the technological breakthrough or innovation, the greater the change that will follow.

Take the case of the airplane. People saw a need to move more quickly across the 3,000-mile continent spanned by the United States. Also, ties to Europe and Asia became increasingly strong. People simply didn't want to spend the time needed to cross the country by train. They also wanted to cross oceans in less time than was needed to travel in ships. There was a readiness, a *demand* for a rapid way to travel. Once the airplane had been invented, the demand kept growing. Bigger, faster, longer-range aircraft were developed.

## PROBLEMS AND CONCERNS

Technology, then, can be seen as a tool. People dream about things they need or want to do. Then, they apply technology to make these dreams come true. The ideas behind the dreams also shape the technologies. Basically, people have concerns for safety, life support, health, comfort, and knowledge. Knowledge underlies all of the other wants and needs. Through knowledge, people meet their needs or wants. As knowledge builds, the ability to develop and apply technology also grows. Consider these elements:

- Safety
- Life support
- Health
- Comfort
- Knowledge.

## Safety

Recall the myth of Icarus, the boy who flew too close to the sun. People's dreams, which become needs and wants, are not necessarily wise. The desire to fight other people is a case in point.

At first, people who lived in any given place had common enemies. Early enemies were mostly natural—fire, flood, cold, wind, and wild animals. As people learned to deal with their natural enemies, they gained the freedom to threaten one other.

**Death and destruction** can result from some types of technology. This photo shows the mushroom-shaped cloud that follows detonation of an atomic bomb. (Courtesy American Petroleum Institute)

Arguments and fights developed as people added to the things they owned. Individuals fought with one another. Groups were formed and increased the size of conflicts. Ultimately, whole nations fought each other. As a result, one of the branches of technology has been the development of weapons of destruction.

Early weapons technologies involved the spear, the bow and arrow, the slingshot, swords, and catapults. Later, there were guns, cannons, then armoured vehicles such as tanks. As wars became bigger, involving more people and larger areas of the world, governments spent more money and effort on weapons technology. The airplane became a threat as well as a means of transportation. The desire for faster and more powerful weapons to fight wars led to the development of jet aircraft.

Ultimately, more advanced aircraft were developed as weapons against the aircraft that already existed. Rockets and missiles became anti-aircraft weapons. Complicated structures were developed to protect people against bombs delivered by airplanes.

Convenience and danger have gone hand in hand as technology has grown. People must use technology with wisdom and caution. The automobile provides an excellent example.

**Both benefits and problems** can result from the same technology. Cars provide transportation. But automobiles eventually become ugly piles of junk that society must handle. (Photo above, courtesy Department of Housing and Urban Development. Photo at right, © Jeffry W. Myers, West Stock)

When the automobile was developed, it met some major wants and needs that people had. People wanted a personal way to get around under their own control. The automobile gave it to them. The United States has been called a nation on wheels. The automobile has also been critical to the development of a number of other countries.

But it is part of the nature of technology that solving one problem seems to lead to others. The automobile brought convenient transportation. It also brought death on the highways. Some 50,000 people are killed in automobile accidents every year. Hundreds of thousands of others are injured.

Throughout the growth of technology, safety and danger have been related. People invented weapons to protect themselves. Then they used weapons on other people. One of the advantages of the automobile was human control. Cars were seen as being easier to handle—and thus safer—than horses. Then, as the number of cars increased, so did human deaths and injuries. Safety motivates new technologies. Then, changes seem to lead to new dangers.

## Life Support

The human body must retain a certain temperature to support life. A healthy, normal body has a temperature of 98.6 degrees Fahrenheit (37 degrees Celsius). This is known as *body temperature.*

Body temperature must be maintained if a person is to continue to live. People who live in warm climates, in areas of the world close to the Equator, need little or no clothing. However, in colder areas, it is necessary to use special aids to maintain body temperature. People do this through the use of *clothing* and *shelter.* In cold areas, people could not maintain body temperature without these *life supports*.

The body itself generates heat to maintain life. The fuels that maintain a body are *food* and water. Thus, food (including water), clothing, and shelter are commonly called the *necessities of life.*

In early times, individuals did what they could to find the necessities they required. Caves were common shelters. Plants provided most of the food. However, from the earliest days, people have been eating the flesh of animals as part of their food supply. In this respect, food and clothing became related. That is, the meat from animals provided food, and the skins provided clothing.

Technology has been applied continuously to make it easier and more convenient for people to provide these necessities. To secure food, for example, people began to cultivate plants. The technology of *agriculture* was born. To help with the work of agriculture (as well as for food), people tamed animals. This technology is called the *domesticating* of animals. Domestic, basically, refers to life in a home environment. Anything relating to activities at home can be described as domestic.

Domestic animals, then, were tame animals that could be kept in or near the home. These animals could be used to carry heavy loads or could be killed for food. These same animals, in turn, provided hides that could be used for clothing. The hides also provided the first movable shelters. Animal hides were joined to form early tents. With these, people gained the ability to move away from caves and other natural shelters. This made it possible for people to travel to areas where they could find better food and climate.

**Domestic animals** have helped to meet the needs and wants of people. Animals have provided both food and labor through the years.

The area of life support, then, shows the connection between types of technology. Some needs—food, clothing, and shelter—could be satisfied from the same source. Animals provided part of the basis for all three.

## Health

At the beginning of this chapter, it is pointed out that human life used to be shorter than it is today. At one time, a 15-year-old person might have been considered middle-aged. That is, people had a *life expectancy* of perhaps 30 years. A life expectancy is, simply, the average amount of time people can expect to live. Before there were doctors and medicines, death came earlier than it does today.

Throughout history, curious people have studied the human body, wanting to broaden their knowledge of how the body works. There have been continuous efforts to discover the causes of illness and to develop cures. Early people undoubtedly

**Technologies for health care** have been developing for thousands of years. Health care is an area of technology that has seen rapid advances.

discovered that, when a body had proper clothing and shelter, people enjoyed better health. Through the years, people discovered ways to set broken bones so that they would heal. Even in ancient times, people experimented with natural medicines. Today's health care workers are finding that some of the natural medicines do work. The use of herbs by the Chinese, for example, goes back many thousands of years.

Ultimately, curious people wished to study the human body more thoroughly. Early healers experimented with surgery to increase their knowledge about how the body works. In modern times, people discovered the use of x-rays to examine the body without cutting into it.

The want is obviously there in the area of health care. People don't want to die. They want to live as long as they can. Neither do people want to be sick. They want to enjoy their lives for as long as they can. Thus, health care (medicine) has been an area of active interest. Improvements have been welcomed. Technology has even found ways to replace parts of the human body.

**The artificial heart** is an example of how scientists and technicians keep trying to improve the technology of health care. (Courtesy Thermo Electron Corporation)

Artificial lungs, kidneys, and hearts, have been used. This branch of technology promises to continue to grow rapidly. Health care has become a major area of human activity, employing millions of people.

## Comfort

The wants that people develop can be described in one word: more. People always want something better or greater than what they have. Thus, as humans found that they could meet their basic life support needs, they began to develop wants. These wants can also be identified as comforts, or luxuries.

Consider what life in a typical city used to be like. During the early 1900s, many American cities began to introduce plumbing into homes. Typically, there would be a supply of gas used for cooking, lighting, and heating. This was a comfort—far easier than going out to buy kerosene or wood for stoves or lamps. When water first came into homes, there was only one pipe, and the water was cold. Sometimes, in apartment buildings, one water spigot served the entire building. People filled buckets with cold water, then heated the water on their stoves to bathe and to wash clothes and dishes. These services seem crude today. But, at the time, they represented high levels of comfort and achievement.

Today, the picture is different. Most people can assume that there will be hot and cold water in their homes. They assume that each home will have at least one bathroom. Homes are lighted by electricity and are served by radios, telephones, and television.

In comparatively recent times, comfort and luxury have been major forces leading to the development of technology. For instance, commercial radio did not begin until the 1920s. Commercial television really didn't get started until the 1940s. This means that your own parents can think back to the time when there was no color television—possibly no television at all. Your grandparents may even be able to remember the days before radio. You may even know some older people who can think back to before 1929—the year movies learned to talk. For the past 100 years or so, technology has been a major fact of life for people all over the world. Progress has been so rapid that technology is often described as explosive.

**Technology for amusement** includes the development of video games and other modern entertainment devices. (Courtesy TRW, Inc.)

**Interest in outer space** dates back to ancient astronomers and is also as modern as space shuttle launching programs. (Illustration at left, The Bettmann Archive, Inc. Photo above, courtesy Florida Division of Tourism)

## Knowledge

Technology has developed more rapidly as people have learned more. *Knowledge* is the basic support, the foundation, of technology. It is basic to the scientific way of doing things that people record their knowledge as they learn. Later scientists learn from those who went before. People keep adding to knowledge with their own study and experience. Ultimately, knowledge has become a technology in and of itself. Special tools, including computers, have been developed and used as knowledge builders.

The technology of knowledge has one special feature: Knowledge is *cumulative.* This means that knowledge builds upon itself and upon knowledge that has been established (accumulated) earlier. Another feature of knowledge as a technology is that it supports, or supplies, all other technologies. Also, the technology for knowledge building fixes, or corrects, itself as accumulation takes place. Items of information considered to be correct today may prove wrong tomorrow. When this happens, a new basis of knowledge is formed.

For some 1,500 years, most people believed the world was flat. A sailor named Columbus, working with the findings of earlier scientists, helped prove that the world is round. The technology of knowledge changed. Future discoveries and future courses of technologies changed on the basis of this knowledge.

**Knowledge about people** was a specialty of Plato, an early Greek philosopher.

**Greek ideas.** Both ideas about knowledge and the content of knowledge have been building for many thousands of years. Early appreciation of knowledge came from ancient Greeks—Socrates, Plato, and Aristotle. Socrates was the first person known to have organized the process of inquiry. The *Socratic method* has come down to the present day as a means of dealing with unknowns. The principle, simply, is to ask a series of organized questions as a basis for building knowledge.

Plato was a student of Socrates. He founded what is believed to be the world's first university. Plato applied the teachings of Socrates to establish knowledge about human behavior. His writings dealt with his dream of creating a *utopia*, or a world of perfection, by reorganizing government structures and organizations.

Aristotle, in turn, was a student of Plato. Aristotle brought logic to the organization of knowledge. He developed systems for classifying and organizing information to build knowledge. In a sense, Aristotle laid the foundations upon which technologies have been built. He did this more than 2,000 years ago.

The ancient Greeks were also responsible for development of an important knowledge-building method—the *delphi process.* This is a technique for the sharing of suggestions or ideas contributed by a number of people.

**The use of logic** was a major contribution of Aristotle, who was a student of Plato. (The Bettmann Archive, Inc.)

Delphi was a Greek town that was the home of the Delphic Oracle. The Delphic Oracle was an idol credited with special wisdom and problem-solving skills. Citizens could ask questions or state problems on slips of paper. These were slipped into a hole in the stone idol. There, they were turned to ashes by a constantly burning fire. The Delphic Oracle was tended by a high priestess and her helpers, priests. The process blended, or mixed together, many problems and ideas. The oracle solved problems by using their combined wisdom. People would enter their questions, then concentrate on their request. During the period of deep concentration, the oracle answered the request by speaking through the high priestess. After the priestess had spoken, the priests added their suggestions or corrections. Thus, the citizen heard several answers.

Today, the delphi process invites ideas and suggestions. Managers or specialists review the suggestions to help solve problems or improve operations. A present-day version of the

**The sharing of knowledge** is a key to the development of science and technology. Modern technology even provides the means to assist this sharing of knowledge by recording and displaying information. (Courtesy Sperry Corporation)

Delphic Oracle is the suggestion box. Businesses use suggestion boxes to invite ideas for improvement from employees or customers.

The delphi process is an old, proven method for sharing and building knowledge. The basic idea is to focus the attention of many people on a problem situation. Applying more thought can often produce a better, more workable plan.

**Scientific notebooks as knowledge builders.** Leonardo da Vinci, the man who first sketched the airplane, was both an artist and a scientist. His paintings are among the most famous in the world. Leonardo kept notebooks and sketchbooks from which generations of scientists have learned how to record and accumulate their own knowledge. His creative mind contributed to the world's knowledge through his art and his scientific writings, which served as a model for future scientists.

**The scientific method.** Sir Francis Bacon, an English scientist who lived from the seventeenth to the eighteenth century, originated what we now call the *scientific method*. Bacon added new approaches to Aristotle's knowledge-building principles. Aristotle had used what is called *deductive* reasoning. This means, simply, that people formed specific ideas on the basis of what was generally seen or thought. These ideas were then recorded. Other scientists, after reading the earlier reports, could take different approaches because the general ideas may have changed. They might change the deductions of earlier knowledge builders, depending upon their own findings.

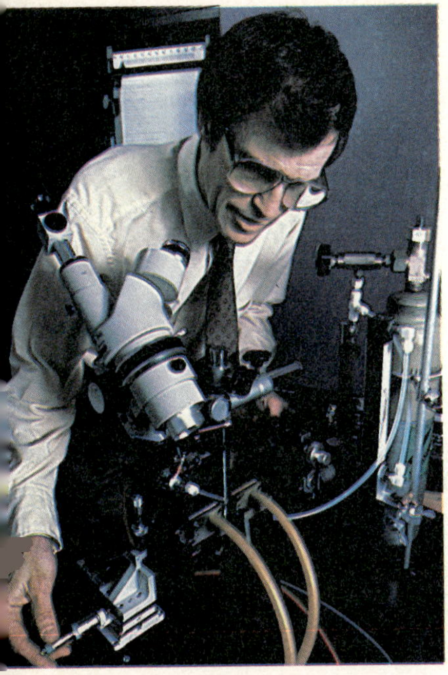

**Experimentation and observation** are elements of the scientific method being practiced in this experiment dealing with fiber optics. (Bell Laboratories)

Bacon added a new dimension. He felt that scientists should test, or challenge, their own observations and deductions. Testing was to be done through experimentation. Under Bacon's approach, a person stated an idea. This statement was known as a *hypothesis*. Once the hypothesis was stated, the scientists performed experiments and reported on the results.

The results of the experiments, in turn, became the basis for scientific knowledge. This is known as the *inductive* approach. Under inductive reasoning, people examine specific ideas to arrive at general ideas. The results of experiments allow future scientists to draw their own evaluations, or *inferences*, from earlier findings.

Sir Isaac Newton was a pioneer in using mathematics for the testing of a scientific hypothesis. Newton is a familiar name to most people. Part of his fame lies in a story that is something of a myth. As the tale goes, Newton was sitting under a tree one day when he noticed an apple falling. He wondered why the apple fell when it left the tree. This led to discovery of the laws of gravity.

Undoubtedly, Newton did see apples falling from trees at some time in his life. But scientific discovery is not all that simple. Long, hard study went into thinking about why things fall and at what rates they are attracted to the earth. In performing these studies, Newton actually invented a new branch of mathematics, called *calculus*. Calculus is a mathematical technique that makes it possible to deal with estimates, or approximations. The scientist gets increasingly closer to the desired results as calculus functions are applied. Newton used calculus to develop the laws of *gravity*, the force that attracts objects in the environment to earth. Gravity affects everything that happens on earth and throughout the universe.

**The law of gravity** was stated first by Sir Isaac Newton.

**The knowledge explosion.** Since Newton's time, science and technology have expanded so rapidly that their growth is described as explosive. There have been many important contributions, but fewer individual discoveries. For example, Thomas Edison made it possible to have electricity in your school, home, and elsewhere. Edison developed the first electric light bulb. He also devised methods for transmitting electricity over long distances through use of wires. But Edison had more than a century of accumulated knowledge about the nature and properties of electricity to draw upon.

As the number of scientists has increased, it has been harder to give credit to individuals. For example, Alexander Graham Bell is recognized widely as the inventor of the telephone. However, many scientists were working toward development of the telephone at the same time. Bell filed his patent application just a week or two before another inventor, Elisha Gray.

In a sense, recent scientists and inventors have been in competition. This situation serves to prove some points about the basic nature of technology. That is, technology serves people. Success in technology follows and meets needs or wants by people. There can be no progress or change in technology until people are ready for new inventions.

Any technology that is successful depends upon people. Further, people are in control of the uses of technologies. Consider something as simple as a walk-around radio-cassette player. This product of technology gained rapid acceptance. Almost overnight, people seemed to sprout walk-around units from their belts and earphones from their ears. But people are still in control. People tune the radios or insert the cassettes.

Another point about the nature of technology relates to knowledge. Knowledge is a necessity that serves all technology.

**Major advances in technology** resulted from the work of Alexander Graham Bell, inventor of the telephone. (Reproduced with permission of A.T.&T.)

## Summary

Technology is the application of knowledge and activity to the needs and wants of people. Needs and wants can be simple, such as prehistoric people's needs for food, shelter, and clothing. Needs and wants also can be complex, such as the urge to explore outer space. Progress occurs when technology is successful in meeting human needs and wants. Technology, then, can be seen as a tool to help people.

Knowledge is both the basis for human wants and needs and for technology. As knowledge builds, the ability to develop and apply technology also grows. Therefore, increasing knowledge has been an important concern for people. Other concerns that have stimulated technological advances include the areas of safety, life support, health, and comfort.

At first, safety from wild animals and other natural dangers concerned people. Later, as people learned to deal with their natural enemies, safety from other people became a problem.

Supplying the basic needs to maintain life has always been important. The necessities of life include food (including water), clothing, and shelter. Technology has been applied continuously to make it easier to provide these life supports.

Health care has become a major area of technology today. People want to enjoy their lives as long as possible. Technology helps people to live healthier, longer lives.

As people found they could satisfy their basic needs, they began to desire comfort in their lives. In recent times, progress in providing luxuries, such as radio and television, could be described as explosive.

Knowledge, the foundation of technology, has become a technology in and of itself. Tools, such as computers, have been developed and used as knowledge builders. As knowledge has increased, so has the rate of invention and technological change.

## Using Technology

1. Based on the information given below, identify whether deductive or inductive methods are being applied. State the reasons for your decision in a short composition. Be sure to describe the difference between deductive and inductive methods.

   *The ivy in the music room had grown noticeably faster than the ivy in the cafeteria. Sam was curious about the difference. He learned from Mr. Esben, the school groundskeeper, that both plants had come from the same cutting. Further, Sam learned that both plants had received equal amounts of sunlight, water, and fertilizer. He wondered if music might have made the difference. At home, Sam decided to test his idea that music affects plant growth. He took two cuttings of equal size from a begonia plant. He put the cuttings into pots of the same size and gave equal amounts of water to both. Sam then put one plant in the kitchen and the other one in his room by the radio. He decided to measure the growth of each plant once a week.*

2. Develop a deduction based upon the set of statements given below.

   - People would like to spend less time shopping.
   - Computers are becoming a common household item.
   - Department stores use computers to keep track of the products they sell.
   - Computer technology continues to develop at a rapid pace.

3. Apply the delphi process to suggest solutions to the problem stated below. Record and evaluate suggestions from at least three persons.

   *Computers are used in most businesses today. Many workers are finding that their jobs require some knowledge about computers. For instance, word processors are replacing typewriters in many offices. Public schools recognize the need to teach students how to use computers. However, most schools do not have enough computers to meet this need.*

## Technology Terms

progress
technology
solution
demand
body temperature
clothing
shelter
life support
food
necessities of life
agriculture
domesticate
life expectancy
knowledge
cumulative
Socratic method
utopia
delphi process
scientific method
deductive
hypothesis
inductive
inference
calculus
gravity

# 2

# TECHNOLOGY AND CHANGE

## Your Learning Job

After you have read this chapter and completed the work assignments, you should be able to:

- ☐ Explain how technology affects where and how people live.
- ☐ Define urbanization and explain its connections with technology.
- ☐ Compare the development of the craft system with the evolution of modern technologies.
- ☐ List and describe five major areas of technology that evolved as cities were built.
- ☐ Explain the role that energy plays in today's technological world.
- ☐ Illustrate how technologies are always changing.
- ☐ Explain why written communication is important to technology.

## CIVILIZATION AND TECHNOLOGY

**The tallest building in the world** is the Sears Tower, Chicago. This building represents an ultimate in urban building technology. (Courtesy Sears, Roebuck and Co.)

The more people there are, and the closer they live together, the more they seem to need technology. Certainly, history shows that the development of technology has taken place side-by-side with *urbanization.* Urban is a word describing anything to do with cities or concentrations of people. Urbanization, as a historic development, describes a movement of people from the land (farms) into cities.

The reasons people moved from farms to cities were also reasons why they needed more technology. In simple terms, there were fewer jobs on farms and more jobs available in cities. Technology was one of the forces behind this change.

Remember the pattern: People first develop a need. Technology meets the need. Then, once the technology is available, changes follow.

A simple way to illustrate this is with supplies of food. Early people learned to grow their own grain so that they would not have to travel continually to find food. The technology of

**Labor saving** resulted when people learned to put animals to work.

agriculture was born. However, farming proved to be hard work. So people began to devise tools to make food growing easier. The plow was invented to make breaking up the ground easier. The scythe was developed to help people cut more grain with less effort. Eventually, people got the idea of using animals—horses and oxen—to pull the plows and cut the grain. The more work animals did, the better life became for people.

Change, in this situation as in others, was a result of technology. As people put animals to work in agriculture, they not only saved labor, but they also grew more food. It was no longer necessary for each family to grow its own food. Instead, farmers had surplus products that they could supply to others in exchange for things that the farmers needed.

The same kind of technological development occurred in other fields. Shoemakers found that they could make more shoes than their own families needed. People could weave more cloth or make more clothing than they needed for themselves. A system of *crafts* resulted. A craft is a special skill that somebody uses to provide goods or services.

Demand for the goods produced by crafts specialists kept growing. The person who made the best shoes could sell the most shoes because people wanted quality. The same was true for cloth, clothing, and other craft items. Eventually, the people

Crafts specialists produced the first goods available for trade. In some parts of the world, crafts are still practiced much as they have been for centuries. (UNRWA)

with craft skills hired others to help them. Their businesses grew. With growth, it became necessary to build bigger places in which to do the work. These centers of production are called *factories*. The concentrations of people who came to work in factories needed housing in nearby areas. Towns and cities resulted. People in these urban areas required additional products and services. Urbanization and technology went hand-in-hand.

Eventually, technology itself followed the same pattern as the crafts. That is, people began to think about and specialize in branches of technology. The same kind of specialization had taken place earlier in such occupations as farming, clothing, toolmaking, and other crafts. Examples of the specialized areas of technology that evolved as cities were built include:

- Production
- Transportation
- Health care
- Structures or buildings
- Communication.

## PRODUCTION

The necessities of life, remember, are food, clothing, and shelter. Cities, in effect, provided a type of shelter, taking care of one necessity. As cities grew, however, demands for food and

**Trade and protection** were the functions of early cities like this one in the Middle East. (Courtesy Jordan Information Bureau)

clothing production changed. The cities provided shelter, but people still had to feed and clothe themselves.

Cities themselves go back many thousands of years. There were towns and cities in ancient Greece, Rome, the Middle East, and the Orient. Cities were formed as hubs of trade and craft activities.

The crafts themselves tended to move into areas close to the materials, or *resources,* they needed. For example, food processing took place close to agricultural growing areas. Clothing factories were located near the factories that wove cloth. More

**An early technology target** was the production of textiles. Hand methods of weaving made cloth costly. Automatic weaving machines made more clothing available, at lower costs, to more people. (© Craig Aurness/West Light)

recently, steel mills were built near the mines that produced iron ore and coal.

There were many technologies and opportunities for people moving from farms to cities. However, all factory workers had common sets of needs. Factory workers had fewer opportunities than farmers to produce their own food or clothing. Other *producers* had to provide food and clothing.

It is understandable, then, that technologies of production developed early. Understanding how food and clothing were produced will help to explain how production methods were developed to meet the needs of people.

The use of animals on farms made it possible to grow more grain. Grains include wheat, corn, rye, and others. These basic grains are ground into a fine powder, or flour. Flour, in turn, is used to make breads and other foods. Eventually, there was more grain available than could be ground efficiently by hand.

The need to grind grain became a challenge for people in that craft—millers. Milling by hand was hard, slow work. As a start toward food technology, animals were put to work grinding grain. You have undoubtedly seen pictures of oxen or horses walking around in circles, turning wheels that grind grain. Even with animals, however, there was not enough production.

**Wind has been a source of energy** for centuries, beginning with sailing ships. Modern wind machines are being used as extra sources of electric energy. (Courtesy U.S. Bureau of Reclamation, Medicine Bow, Wyoming)

## Energy and Power

What was needed to produce more food was technologies using *energy*. One of the first technologies built around energy, other than animal or human labor, was based upon the wind. Windmills were constructed. Long sails, or paddles, caused the grinding wheels to turn and produce flour from grain. This process illustrates one of the basic principles, or rules, of technology. That is, energy is converted to *power*. Energy is a source of power. Power, in turn, is the rate at which work can be done by applying energy.

Wind is a natural source of energy. That is, wind exists in nature. People recognized its *potential* for work and devised technologies to apply, or harness, the energy of the wind. Energy that isn't used for power can be called *potential energy*.

Another important natural source of energy is running or falling water. The weight of moving water generates a *force* that produces energy. A force is the use of energy in a specific direction or type of movement. For example, when water is free to move, it will always fall or drop downhill, following the laws of gravity. The force of water can be seen in any river or stream. These are natural forms of force that direct the flow of water—always downhill and in specific directions. The force of flowing water can be applied to production by placing circular wheels with paddles into the flowing water. These paddles capture the potential energy and turn it into some form of power. In the modern world, water wheels generate electricity. In older times, water became a critical source of energy for production. In agriculture, water wheels were used to provide power for grinding grain.

Energy and power also were important parts of the development of technology in making clothing. The main source of clothing for the past several centuries has been cloth. Cloth, in turn, is woven from thread or yarn. The weaving process involves the interlacing of thread in crossing patterns so that the thread holds together to form cloth. Historically, cloth has come from two main sources: wool and cotton. From both of these *raw materials,* thread or yarn to make cloth is created by a process known as spinning. Cotton or wool strands, called fibers, are turned continuously. These fibers attach themselves to other fibers, creating a continuous length of thread or yarn wound around a long stick called a spindle.

**Water was one of the main** power sources for factories during the Industrial Revolution. Streams or rivers caused the waterwheels to turn, transferring power to factory equipment. (© Craig Aurness/West Light)

For many centuries, thread or yarn was formed on devices known as spinning wheels. The person using a spinning wheel either turned a crank or stepped on a foot pedal (treadle). The turning motion resulted in the spinning of thread or yarn.

## Power Sources Attract Factories

To meet growing demands for clothing, water power was used to drive machines that spun thread automatically. A single machine could wind new thread onto a number of spindles. Then, as thread became available, water power was used to drive looms that wove cloth. The spinning of thread and weaving of cloth form the *textile* technology. Textiles was one of the major technologies that led to the building of a number of cities in England.

When the United States was founded and began to build, the textile industry was one of the main job sources in the New England region. The New England textile mills produced thread and cloth from cotton grown in the South.

For centuries, water was the dominant source of power for the production of thread, cloth, and many other items that people needed. After a time, there weren't enough good locations beside enough running streams and rivers to provide all the power that was needed. Also, industries wanted to apply production technologies in places that didn't have running water. To meet this demand, scientists and engineers developed the steam engine.

Steam engines burn fuel to boil water. The steam from the boiling water expands, or takes more space. The expanding steam provides force that is turned into power. With a source of energy that could be located right in a factory building, it became possible to set up production anywhere. The steam engine became a major force for growth and change. Again, technology met a demand. Once technology happened, change followed.

In time, the steam engine was replaced by a cleaner, lower-cost source of energy—electric motors. That is, electricity provided energy and motors provided power. Today, most of the things you eat or use were produced with the help of electric energy.

**The steam engine** proved a major source of power for the spreading of the Industrial Revolution. In this historic drawing, a steam engine is used to ease the work of loading cargo onto ships.

## TRANSPORTATION

As explained above, many factories were located near the sources of materials they needed. Factories that made glass had to be in places where they could find silica sand, the raw material for glass. Steel mills had to be near iron ore and coal mines. In England and the United States, textile mills had to be near ocean ports. This was because cotton was usually transported on ships. The cotton used in England came from such places as Egypt, India, or the American South. The cotton for the mills in New England also came by ship from the American South.

For both raw materials and the sale of items already produced, faster forms of *transportation* had to be developed. People have always had a need to transport themselves and the things they own. To transport means to move from one place to another. Transportation also assumes that there is a *point of origin* and a *destination.* That is, transportation includes getting from a known starting point to a known arrival location.

In the earliest years, people simply carried what they owned in their hands or on their backs or heads. Thus, a cave dweller would scout nearby forests for sticks and carry them back for the fire. As the population grew and fires became larger, animals were used to carry the sticks. Animals were also used to carry the things that belonged to people, such as their animal-skin tents. As people traveled farther, they discovered that riding animals was both faster and easier than walking. Eventually, people built *vehicles* that could be drawn by animals. A vehicle is any device that can move people or things. Some vehicles have wheels. Other vehicles fly. All vehicles need sources of power.

**Industrialization required transportation** that would be free of the limitations of waterways. Railroads were the transportation method for building the United States. Railroads remain a main source for carrying freight between cities. (Association of American Railroads)

Another major basis for transportation has been water. Very early, people discovered that certain materials, such as wood, float on water. They built rafts, then canoes, then ships. At first, people used only natural waterways—rivers, lakes, and oceans. When there was no natural waterway, the convenience of floating was so great that people built their own waterways. There is evidence that the Chinese dug canals for their transportation more than 2,500 years ago. Many of the world's major cities or areas have been served by canals. The city of Venice, Italy, used canals as its major form of transportation. Early Mexican cities had canals long before the Europeans arrived. In the United States, the Erie Canal, serving New York State, was a

**People have traveled a long way** since the days when they depended upon their own feet as their chief means of transportation. Progress has come through the domestication of animals, through the invention of cars and trucks, along with the building of roads, and through the use of rivers and canals for travel by ship. (Credits: above, left, UNRWA; above, right, © Jim Scruggs/Uniphoto; right, New Jersey Division of Travel and Tourism; lower, right, photograph by the Panama Canal Commission)

**Trains and planes** have played important roles in the evolution of modern transportation. Air transportation has progressed from crude, two-passenger planes to the jet age with complex airports. Trains have gone from steam engines, to diesel-electric locomotives, to modern bullet trains like those at left, now used in Japan.

(Credits: above, left, American Airlines Photo; right, Grumman Corporation; center, Santa Fe Railway Photograph; below, © 1981 David Ryan/Uniphoto.)

major factor in settlement and growth. Certainly, the Suez Canal and Panama Canal were important in making world travel faster and more convenient.

On land, people built their transportation technology in the form of roads, bridges, railroads, and streetcar lines. Until the early 1920s, the main source of power for vehicles used on land was the horse. Horses pulled streetcars, carriages, wagons, and barges on canals. People also rode horses in transporting

**Waterways were early highways** for the transportation of people, equipment, and materials.

themselves and their belongings. People also developed a series of *engines* that could power their land vehicles. Engines are machines that change energy into force. People invented the steam engine and the internal combustion engine. Steam is a source of power because heated water expands into a gas. The force of the expanding gas, in turn, can generate power. Steam is the source of power for an *external combustion* engine. That is, the heat that produces the energy of expanding gas is outside the container holding the water. External means outside.

Another type of engine, the main force for transportation today, uses *internal combustion.* This means, that burning of a fuel (combustion) takes place inside the engine. For example, in most automobile engines, gasoline is burned, or combusted, inside engine compartments called *cylinders.* The burning causes the gas to expand, thus driving pistons that convert the energy of combustion into power by turning a drive wheel.

Internal combustion engines have been used to replace other sources of power on water, as well as on land. For example, boats and rafts used to be moved by poles stuck into the bottom of a river or canal. At other times, oars were used. For many thousands of years, wind was the main source of energy for ships. The energy of the wind was converted to power when it was captured in sails.

The replacement of one form with another is part of the nature of technology. Many technologies have existed for a long time, as though they were waiting for somebody to invent

**Steam-powered ships** proved an important technology for moving goods across oceans. Steamships improved delivery schedules for raw materials and manufactured goods.

something better. Thousands of years ago, sails were a major improvement over oars. Then, the steam engine was invented. People could cross the Atlantic Ocean in, perhaps, two or three weeks instead of as much as two months. Within a few decades, steam replaced sails as the main source of power for ships. Just as quickly, perhaps faster, steam was replaced by the internal combustion engine. Now, nuclear-powered submarines and ships are common.

Air travel, as described in the previous chapter, had been a dream of people for many centuries. People were ready for the experience and the convenience of air travel. Even before the airplane was invented, people were flying in balloons and in dirigibles. Both of these are known as lighter-than-air craft. For example, a balloon weighs less than the air that occupies the same space. Therefore, just as a ship floats on water, a balloon floats on air. The same is true for a dirigible. The differences are that a dirigible has a source of power, usually a motor with a propeller, and can be steered. A balloon, by contrast, floats where the wind takes it.

**A long-standing dream came true** when people developed a technology for powered flight. This historic photo shows a Wright Brothers airplane in 1903. (The Bettmann Archive, Inc.)

The airplane offers a dramatic illustration of how fast technology can develop once demand and knowledge exist. The first successful commercial airliner, the DC-3, was introduced in 1936. By the late 1940s, air transportation was the main method for moving people over long distances. By 1959, passengers were flying in jet-propelled airplanes. Now, compare this progress with the use of sails as the main source of transportation power. Sails were around for 30 or 40 centuries.

## HEALTH

As far as the health of people is concerned, cities have meant good news and bad news. Originally, cities had several advantages. People had protected places to live. They were sheltered from the weather. Also, they were protected from wild animals and other dangers of the countryside.

However, cities also helped to demonstrate that people can be bad for one another's health. With people living and working closely together, packed into small areas, diseases began to develop and spread. The distance and the fresh air of the countryside that helped to provide some protection were gone in the cities. Sickness and death became realities of everyday life. Diseases and plagues spread rapidly through cities, causing thousands of deaths. A plague occurs when an infection or disease spreads out of control. During one plague in the fourteenth century, half of the working people in the city of Florence,

Italy, died. The killer was the disease known as the bubonic plague, or the black death. This disease killed millions of people over hundreds of years before its cause was found. There is still no cure. But understanding the cause made it possible to apply controls.

The plague stems from rats who come to cities to live on garbage. Fleas, which infest the coats of rats, pick up germs and carry them to people. The flea bites transmit the plague. Understanding the cause has helped to apply measures for controlling rats and fleas. However, many thousands of people around the world are still struck by the black death every year. Of these, more than 25 percent die.

**Health care took strange forms** during the seventeenth century. This outfit, developed in Germany, was designed to protect its wearer against the plague. (The Bettmann Archive, Inc.)

For thousands of years, frequent sickness and early death were accepted as inevitable parts of city life. Diseases and plagues were thought to come from supernatural causes about which people could do nothing. Then, during the late nineteenth century, scientists reasoned that diseases must be caused by germs. From that point on, human life has grown longer and more productive.

Among the important discoveries of science that has affected human health is the importance of cleanliness. In early cities, sewers and garbage dumps became breeding grounds for diseases. Once the causes of disease were known, systems of sewers and other forms of waste disposal were built.

Within a period of perhaps 30 years toward the end of the nineteenth century, many key discoveries were made. These included the germ theory of diseases, the control of bacteria through pasteurization, and the x-ray. X-ray pictures provided internal pictures that greatly improved the ability of physicians (doctors) to make people well. Many illnesses that used to kill people became curable. Extra years of useful, enjoyable life have been made possible for the average person. A century ago, health care was virtually unknown as a science. Today, it is a major technology.

## STRUCTURES

Since people began to move into towns and cities thousands of years ago, structures have been a problem. At first, people simply built structures they wanted on whatever land they could find or buy. It wasn't long before a need was recognized to get

**The infrastructure of modern cities** includes all services required by residents, including fire fighting equipment and the streets and roads that handle traffic. (Both photos, Queen City Metro, Cincinnati, Ohio)

things organized. Streets, sewers, streetcar lines, railroads, and other services were needed.

All of these key services, taken together, are known as the *infrastructure* of a city or area. To provide the facilities and services of an infrastructure, cities had to pass laws about buildings. Buildings could only be built in certain areas. Each structure needed specific features to provide for life support of the people who lived or worked in it.

These rules or laws about buildings are called building *ordinances.* Ordinances are designed primarily to assure that the health and safety of people who live in an area are protected. For example, ordinances require that buildings occupied for living or working must have fire escapes. There must be a source of fresh air in living and working areas. Any fuels used in buildings must be handled safely.

All of these ordinances are the result of technology. So are the *zoning ordinances*, which set up rules about where buildings can be constructed and under what conditions.

As cities grew, there were more people than there was room to put them. At first, cities began to spread out. Then, the homes and factories were so far away from the central points in the city that travel time was too long. People who lived and worked in cities experienced a new need: They required safe, clean living and working quarters that were more compact. Putting it

**Density of housing** has been a characteristic of construction in cities. Before the elevator made possible high-rise construction, low buildings were constructed in tightly packed rows. (© Les Moore/Uniphoto)

another way, the *population density* had to be increased. That is, it was necessary to make it possible for more people to live and work in smaller land areas.

In terms of technology, buildings are called *structures.* The technologies for creating structures are known as *structural design* and *construction.* The people who design structures are called *architects.* The people who figure out how to build the structures designed by architects are called *engineers.*

As cities grew, the challenge to architects and engineers was to create taller buildings. This appeared to be the answer for increasing the population density.

One of the first structures to prove that tall buildings could work was the Eiffel Tower, completed in Paris in 1889. Around

**Some of the oldest and some of the newest** designs for skyscrapers appear in this photograph of Chicago. (Courtesy Sears, Roebuck and Co.)

the same time, Chicago went through a major reconstruction following a disastrous fire that destroyed most of its downtown area. The rebuilding of Chicago attracted architects and engineers from all over the world. One of these people, Louis Sullivan, is recognized as the father of modern architecture.

During the 1880s and 1890s, Sullivan developed the structural principles for the skyscraper. Sullivan's work depended on another technology that made change possible. This technology was the elevator. A steam-driven elevator had been invented by Elisha Otis in 1861. The powering of elevators by electricity during the late nineteenth century made it possible for Sullivan to design and build the first skyscrapers. A new type of city was on the way.

## COMMUNICATION

All technology depends upon knowledge. Knowledge is an accumulation of the experiences and creations of people. To share their knowledge, people need to be able to communicate.

In the oldest form of communication, people simply speak to one another, face to face. However, knowledge about technology has to be shared beyond face-to-face contact. Also, communication about technology can be complex. People need to be able to explain themselves in writing. As needs for communication increased, people developed alphabets from which they could build their words. (The word alphabet, itself, comes from the first two letters of early written alphabets. The first letters of the Hebrew alphabet are *aleph* and *bet*. In Greek, the first two letters are *alpha* and *beta*.)

Other inventions that made it possible to use written language for communication included paper, writing instruments, and ink. Early uses of written languages included commerce and religion. Written languages are the basis for the building and sharing of knowledge. Therefore, the technologies of writing had to be in place before science and advanced technologies could be developed.

The communication field also demonstrates the rate at which the growth of technology has accelerated. A major technology that made mass communication possible was the invention of moveable type by Johann Gutenberg. This took place around 1450. Before this time, books were written or copied by

**Movable type** made it possible to assemble sets of letters or numbers like those shown above into complete books or other publications. An early printing shop using movable type is shown in the historic drawing at the left.

hand. The preparation of a single volume could take months. Only the most wealthy persons could afford books. Gutenberg developed a way to make moveable type by pouring hot metal into molds. These units of type could then be lined up in a printing form. To make impressions on paper, Gutenberg adapted a tool that was handy at the time, a wine press. This is why, through the years, printing devices have been called presses.

After Gutenberg's invention, the forming of messages for printing—*typesetting*—remained pretty much the same for more than 400 years. The next major change in this technology came in 1884, when Ottmar Mergenthaler invented a machine known as a Linotype. The Linotype machine made it possible to use a keyboard to select the letters and words to be set into type. The machine then cast a whole line of type at a time. This sped production considerably, as compared with picking single letters by hand.

More recently, computers have been put to work on typesetting. On modern newspapers, for example, reporters write their stories directly on terminals—machines attached to computers. The computer then drives a typesetting machine that can produce a complete page of a newspaper in less than two minutes.

**Information technology** has moved ahead rapidly with the introduction of computers that process data, deliver information, and even serve as learning tools. This photo shows the use of a small computer for student instruction.

Here is another example of how technology accelerates change. Gutenberg's method dominated printed communication for more than 400 years. The Linotype was the main machine for perhaps 70 years. Then, within a period of less than five years, computerized typesetting took over. Almost all information published on paper today is processed through computers at some point.

Gradually, printing presses also advanced their technologies. Earlier models were hand-powered, then steam driven, now electrical. Today, newspapers in large cities are printed at rates of between 70,000 and 80,000 copies per hour.

Another example of the rate at which communication technology has exploded can be seen in the case of the office copier. The first xerographic printing device that made dry copies in business offices was introduced in 1959. Since then, copy production has become a worldwide industry, transacting tens of billions of dollars of business each year.

The computer itself is another example of the speed with which technology can take hold. The first computer to be used for a practical job was installed in the United States Bureau of Census. It was used to process the 1950 census data. The first computer to be used commercially was installed by General Electric in Louisville, Kentucky, in 1951. This computer was used entirely to process payrolls. A few years earlier, one of the world's leading business machine companies decided not to make computers. The company commissioned a study to determine the potential market for computers. The report predicted that the future demand for computers would be a total of 300 units. Today, computers are being manufactured at a rate of more than 2 million annually.

A technology, any technology, once it is introduced, seems to develop an identity and a force of its own. Once a company has a computer, for example, it follows almost automatically that people will find more jobs to be processed. Once a copier is placed in an office, people find more things that need copying. Thus, while technology fills needs, technology also invites use.

Today, technology is a major force in the lives of almost everyone. Technology seems to be a force without limits. But there are definite limits. Technology uses resources. These resources, in turn, exist in limited supply. Thus, in learning about technology, it is also important to understand the limits that exist. Only so many resources are available in the world. These resources and their relationship to technologies are described in the chapter that follows.

## Summary

Technology brings change to people's lives. For example, people moved from farms to cities as a result of technology. Technology eliminated jobs on farms and created jobs in cities. As cities grew, the way people lived changed. New needs emerged, and new technologies were developed to meet those needs.

Areas of technology that developed as cities grew include: production, transportation, health care, structures, and communication.

The technology of production included the development of sources of energy. Early sources of energy were moving water in streams and rivers. Later, people developed external and internal combustion engines. In time, electricity became the main source of energy.

As methods of production improved, attention turned to the transportation of people and goods. The need was to move more people and goods in less time. An early source of power for transportation was the wind. Sails powered ships for thousands of years. Then, within a few decades, steam replaced sails as a main source of power.

Urbanization created conditions that brought plagues and diseases to cities. Therefore, better health care was needed. Eventually, people realized the importance of cleanliness and sanitary conditions.

As more people moved to cities, structures that were both safe and compact were needed. The technology of structural design answered the need for more compact design with taller buildings. To protect people's health and safety—and for efficient use of land—ordinances were passed. These ordinances covered the location and design features of buildings.

Technology depends upon accumulated knowledge, which, in turn, depends upon written communication. Thus, the technology of written communication has been essential to all other technologies. The technology of communication has advanced from the invention of ink and paper to the computer.

## Using Technology

1. Technology affects the way people live—how they dress, prepare food, move from place to place, even how they play. Recent technology has affected many popular games and created some new ones. Write a short description of these changes.

2. Recall the pattern of technology: People develop a need. Technology meets the need. Then, once the technology is available, change follows. One such development has been television. Think about what your life would be like without television. In a short composition or discussion in class, describe the major changes in modern life that followed the introduction of the technology of television.

3. Technology affects everyone. Older citizens can recall times before certain technological developments that you are used to. Interview someone who is many years older than you. Ask about changes that have occurred in his or her life due to technology. Jot down the answers and share your findings with the class.

## Technology Terms

urbanization
craft
factory
resource
producer
energy
power
potential
potential energy
force
raw material
textile
transportation
point of origin
destination
vehicle
engine
external combustion
internal combustion
cylinder
infrastructure
ordinance
zoning ordinance
population density
structure
structural design
construction
architect
engineer
typesetting

# 3

# HUMAN AND NATURAL RESOURCES

### Your Learning Job

After you have read this chapter and completed the work assignments, you should be able to:

- ☐ Define the phrase human resources and describe the roles of people in relation to technology.
- ☐ Define the term demand and tell what factors influence the demand for technology.
- ☐ Explain the importance of invention to technology.
- ☐ Trace the steps taken to implement a technology.
- ☐ Explain how control of a technology relates to supply and demand for needed resources.
- ☐ Define the term materials and explain how these natural resources are used in technology.
- ☐ List several sources of energy and explain their importance to technology.
- ☐ Explain how resource supplies can affect people and technology.
- ☐ Explain and illustrate the idea of a system of technology.

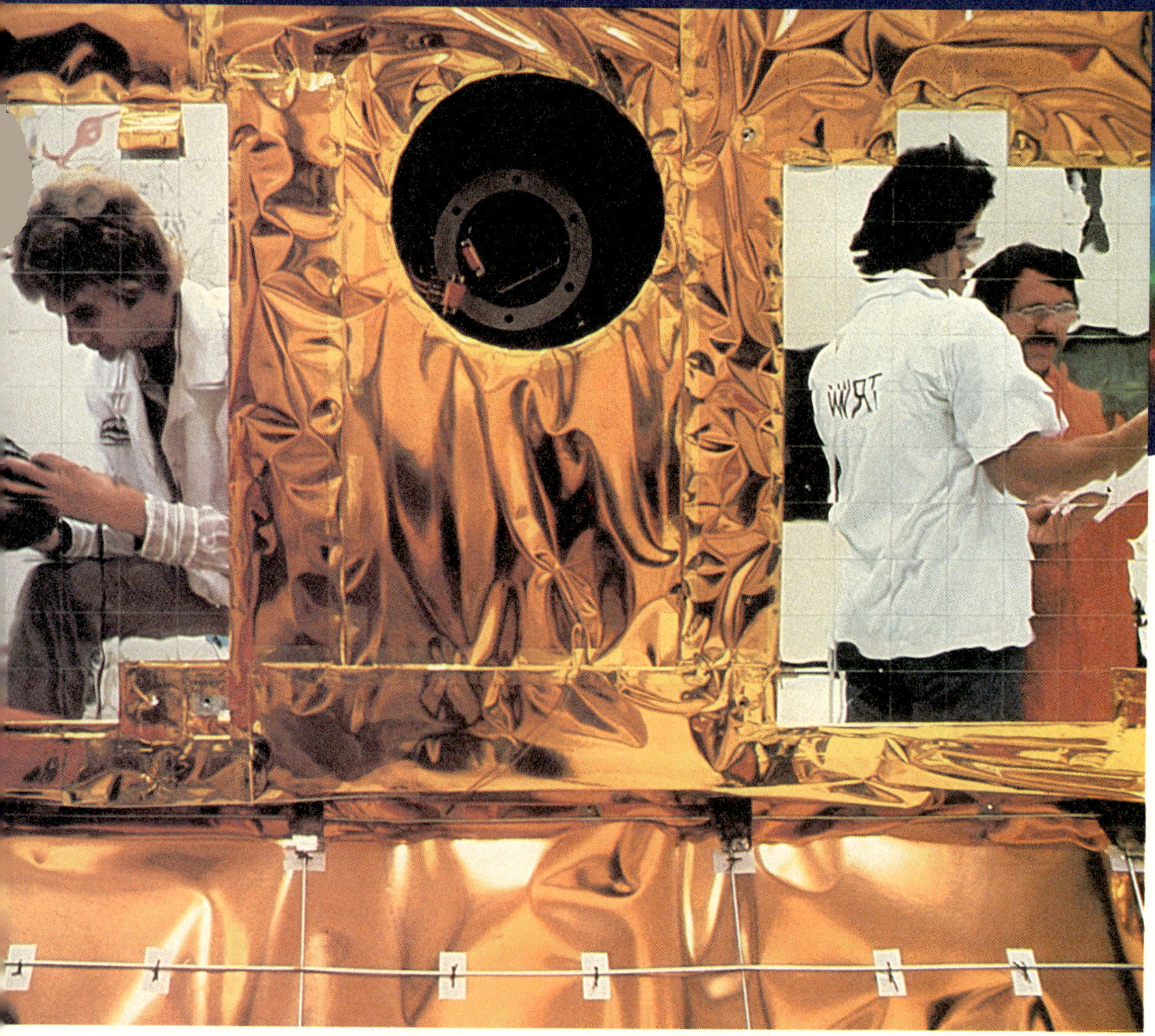

## THE VITAL IMPORTANCE OF RESOURCES

Modern life depends upon technology. Technology meets both needs and wants for most people. In turn, technology presents its own needs and wants. That is, technology depends upon certain human and natural resources.

*Human resources* include all the elements that people bring to the development of end products of technologies.

*Natural resources* are materials or energy sources that are found in nature. That is, natural resources are part of the Earth and its environment.

**Human resources** are the most important basis for technology. No matter what kind of equipment may be developed, people are still the key to the design and use of new technologies.

(Courtesy TRW, Inc.)

**Oil consumption** for the typical family of four in the United States averages 180 barrels per year. (Courtesy Standard Oil Company of California/The American Petroleum Institute)

## HUMAN RESOURCES AND TECHNOLOGY

Technology is for people. People perform a number of important roles in connection with technology. These include:

- Demand
- Invention
- Implementation
- Control.

### Demand

*Demand* simply means that people are ready to accept a new technology or the products of an existing technology. Demand takes in both needs and wants that people identify. Needs, of course, are the things that people require for life support. Wants are related to comforts or conveniences.

Demand changes as people experience changes in their environment or in their everyday lives. A good example can be seen in how people meet their needs for housing and transportation. In 1973, the cost of petroleum products more than doubled. Then, the cost of these products doubled again by 1979. Thus, in one decade, the price of gasoline went from approximately 30 cents a gallon to more than $1.

*Petroleum products* are all items that come from the handling of *crude oil.* Included are gasoline, heating oil, and aviation fuel. Also included are most plastics products. Further, many of the synthetic (artificial) fibers used for clothing and home furnishings are made from petroleum. Examples of artificial fibers include nylon and many other similar fabrics.

As long as petroleum was inexpensive, people were perfectly happy to use petroleum products without concern. When the price skyrocketed, however, new demands were created. People began to explore the potential of solar energy to heat their homes. Many people used fireplaces for heat. Previously, the same fireplaces were considered largely as decorations. Other people even installed wood-burning stoves in their homes.

A new demand in automobiles occurred also. Until 1973, the tastes of Americans in automobiles were for the large and powerful models. After 1973, people began demanding smaller cars that would go further on a gallon of gasoline.

These developments say something about the nature of demand for technology. That is, needs and wants in the modern world usually reflect a number of factors. Life support and basic

**The main source of energy** for the modern world is oil pumped from the ground by devices like the one shown here. (Courtesy Mobil Oil Corporation/The American Petroleum Institute)

comforts form only part of the picture. In addition, cost and availability represent other, related factors. Cost, simply, is an indication of how much has to be paid for a need or want. In the United States, costs are measured in dollars. In Japan, costs are in yen. In Italy, the basic item of money is the lira; in Mexico, the peso. Throughout the world, each country has its own system of money, or *monetary system*.

Money, wherever it is found, is known as a *medium of exchange*. That is, money is the way that people deal with one another to get the things they need and want. A worker sells skills and time for wages, paid in money. These wages, in turn, are used to buy food, clothing, and shelter. Then, anything left over can be used for wants. All told, the exchanges that take place for the satisfying of needs and wants are known as the *economy*. An economy can cover an area, country, or even the world. All of the factors that make up an economy are known as an *economic system*.

Demands for technology, then, are shaped by economic factors, as well as basic needs and wants.

## Invention

*Invention* is a mental process for putting ideas to work. Any specific technology begins with an idea for a better way to do something. Invention lies in the minds of people. People use their intelligence to figure out how to make ideas come true.

Invention is also the beginning of any technology. That is, before a technology can be developed and used, it must exist as an idea. Technologies are brought into existence by inventions. Some of the great inventions that have changed the lives of people are discussed in earlier chapters. These include the wheel, which made it easier to travel, the use of fire, and the invention of engines.

**Implementation of a technology** can have important effects upon the way products are produced and used. This photo shows use of robotic equipment in automatic manufacturing. (Courtesy TRW, Inc.)

## Implementation

*Implementation* of a technology includes all of the steps needed to make something happen. Once a technology is invented, the work of changing people's lives really begins. People must be trained to build the new products. Materials to make the products must be found and delivered to the producers. Factories must be built. Finally, the results of all these efforts, the end product, must be made available to users.

The use of petroleum for a number of technologies will illustrate the process of implementation. The petroleum industry has brought changes to every part of the United States. Throughout the country, oil wells have been sunk wherever pools of oil have been discovered under the earth. There are even hundreds of oil wells operating from platforms over the oceans. In the United States itself, use of crude oil for petroleum products has grown to some 9 to 10 million barrels per day. Each barrel of oil contains 42 gallons.

Long, massive pipelines have been built to move oil directly from wells to *refineries*. A refinery is a place where crude oil is processed to form gasoline, fuel oil, and other products. Other pipelines connect refineries to storage tanks. Trucks and railroads deliver oil products from these large storage tanks to points of use. Many billions of dollars have been spent building these pipelines and storage facilities. Implementation of systems to produce and deliver petroleum products form a major part of the world's economy. In other words, the world's largest businesses are oil companies.

**Natural resources can be used rapidly** as technology advances. A major example is the Alaska pipeline, right, which carries millions of barrels of oil more than 800 miles to the port of Valdez, above. From there, the crude oil is shipped to refineries on the West Coast of the United States. (Alyeska Pipeline Service Company)

## Control

Technology meets demands of people. As technology has developed, demands have increased faster than supplies of technological products. This is one of the basic principles of any economy. That is, people always want more products than are available. For this reason, economics has been called the science of managing shortages.

Because demands exceed supplies, society needs a series of ways to control the distribution and use of the products of technology. Controls over products of technology run in cycles.

When something new is introduced, there may be conditions of temporary surplus. A *surplus* exists when there are more products available than people want to buy. When a technology is new, people may not know about it even though the demand is there. Therefore, one of the human resources used in the control of technology is *marketing*. Marketing is the effort by people to describe a product, offer it for sale, and deliver the products to customers. When a new product of technology is available, marketing people offer it to those who may profit from its use.

**Special marketplaces** develop for each technology or industry. At trade shows like the one illustrated here, hundreds of manufacturers display new products for thousands of visitors. Thousands of shows of this type take place every year.

As a product or technology comes into use, shortages or other problems can develop. One way that such shortages affect people is through price increases. Take gold as an example. Gold is mined from the ground and produced in limited quantities. The demand for gold is always high because gold retains a value. People use gold for jewelry. They also save gold in much the same way as money is put into a bank. In recent years, the price of gold in the United States has increased more than 10 times. Before the increase started, only the government owned gold. Americans were not allowed to own gold personally. Once this restriction was removed, people began to buy and save gold. Under the restrictive system, the price of gold was $35 per ounce. With the restrictions removed, the price has soared to as high as $800 per ounce at one point. When the price of gold reached that point, people didn't buy as much. This is one way, then, in which the resources of technology are controlled. That is, the price changes until it reaches a point at which demand and supply are balanced.

**An example of a regulated industry** can be seen in electric power companies. All phases of this industry, including generating, distributing, and delivering electricity, are controlled by regulatory agencies. (Union Electric Co./American Petroleum Institute)

Another method of control is through *legislation*, or laws, covering the use of technology and resources. For example, the public utilities serve all people in an area and are controlled by the government. Utilities provide services such as telephone, electricity, and gas for heating. In each area, one organization, or company, is assigned to provide a specific utility service. The idea is that this approach is more efficient. It would be confusing to have multiple companies digging up streets to put in gas lines, telephone cables, or electrical wires. On the other hand, with a single company controlling an entire area, supply and demand could get out of hand. The company might charge extremely high prices because people would have no place else to buy the service. To make sure this doesn't happen, the government *regulates* supplies and prices. The government watches the operations of the utility companies and sets a reasonable price for the service. Regulation provides still another method of control.

Regulation is also used when there is a question about the safety or supply of resources. One example can be seen in the use of land within cities and states. Land is a critical, vital resource. There is only so much land in the entire world on which to grow food or construct buildings. As areas become urbanized, the demand for land to be used for factories and houses increases. The uses must be balanced with the needs of all the people.

Somebody, clearly, must decide where the streets will be built and where buildings will be located within the network of streets. In a city, land use is controlled through *zoning laws*. A zone is any area that can be described or identified specifically. A zoning law tells people what kind of buildings can be constructed at which locations. Similar types of regulations also control other important resources.

Still other resources involve the *environment* in which people live. An environment is a specific area. An environment includes everything within, above, and below the area. Thus, an environment includes the ground within the area, the water at ground level or below the ground, the buildings, the plant life, and the climate. If any of these elements of an environment are threatened, regulations can be passed to apply controls. For example, in many areas, the burning of fuels in factories and automobile engines causes chemicals to be discharged into the air. These chemicals may be unhealthy for people. The result is air *pollution*. Polluted air can damage or kill plants and trees. As people began to understand the dangers of pollution, they passed laws to control the use of dangerous materials. These laws are designed to provide *environmental protection*.

**The need for balance** in the use of natural resources is demonstrated in the two photos on this page. The need for oil justifies drilling on the ocean floor. Offshore drilling, in turn, holds the potential danger that oil will spill and will damage life in the sea, as has happened to the seals at right, who have died as a result of an oil spill. (Both photos, courtesy Sun Company, Inc.)

The point is that technology can have both good and bad effects. Human resources play a critical role in helping people to benefit from technology while avoiding the problems. Putting technology to work is the job of people, their governments, and the organizations that serve them.

## NATURAL RESOURCES AND TECHNOLOGY

The needs met by technology are primarily food, clothing, and shelter. Technology meets these needs by putting natural resources to work. It is impossible, in any one book, to deal with every natural resource. However, an understanding of natural resources is important. Natural resources hold the key to the development of technologies. Technologies, in turn, shape the way people live and work. As a future working citizen, you should understand the basics of technology. These basics include the discovery and potential use of natural resources. Natural resources also control technology. That is, as natural resources are used up, technologies must change.

**A natural resource in abundant supply** is coal. Coal reserves in the United States, such as this mine in Wyoming, are believed to hold enough of this raw material to last for centuries. (Kerr McGee Corp./American Petroleum Institute)

For the purposes of this general review, natural resources can be broken into two broad categories:

- Materials
- Sources of energy.

*Materials* include all substances that can be used to produce the products of technology. Materials include solids, liquids, and gases. The air you breathe is a material used by technology. So is the ground on which you walk and the water you drink. Materials are used to build structures or devices that deliver power.

*Sources of energy* take in all forces or potential forces that can be used to produce energy. Some of these sources of energy—such as wind, running water, and the sun—are part of the everyday environment. Other natural sources must be dug from the earth. Included in this category are crude oil, gas, coal, and other materials that can be burned.

Materials and energy are related to each other as natural resources. In some cases, technologies use energy to process materials. In other cases, materials become the source of energy. This is part of the way things are in the world of technology. That is, technologies relate to each other.

## Materials

Materials used by the processes of technology are sometimes called *raw materials*. Raw materials are the basic ingredients used by technology. Raw materials are changed, or transformed, into other products.

As an example of how materials are transformed and used, consider the buildings and streets in your area. One of the materials you may notice is stone. Stones for building can be picked up from the ground or cut from rock formations. Bricks, made from lime and clay, may be used. To be used in buildings, these materials must be held together by cement. Often, steel rods are used to reinforce the walls of buildings. Other materials can include glass, made from sand, and concrete, produced by combining gravel and cement. Steel, made from iron, is another common building material. Copper is used in electric wires and for water pipes. Wood is another major building material.

Different types of materials are used in the technology of agriculture—the growing of food. These include important natural resources, including soil, sun, and water. In addition, a number of resources produced by technological processes also support agriculture. These include fertilizers and chemical insect and weed killers.

The technologies for producing these materials can be related closely. For example, agricultural technology is involved in growing trees used to build homes. Agricultural technology also grows certain fibers, such as cotton, that are used for clothing. Other agricultural products are used to produce fuels used as sources of energy.

**Resources that can be replaced** include agricultural products. These products are harvested regularly, then regrown. (USDA Photo)

**Direct relationships** often exist between raw materials and end products of technology. This is the case with cotton, an agricultural product, left, often used in the production of clothing or textiles. (Photo at left, USDA Photo)

Of course, the main purpose of agriculture is to produce food. This is a vital function as the population of the world continues to grow. In this respect, agriculture depends on another technology, transportation. For people to use food, it must be transported from where it is grown to where it will be eaten.

## Sources of Energy

Some of the natural sources of energy have already been described. These sources include the wind that turns the vanes of windmills and powers sailing ships. Also included were running water and the sun.

Most discussions of the sun as a source of energy, known as *solar energy,* deal with natural heat. The sun can either heat buildings with its direct rays or heat stored water. In addition, however, the sun provides other potential sources of energy. The sun's heat, for example, is a source of energy for a heat pump. Heat is pumped out of an area, leaving cold air. Thus, the sun's heat can be used as a source of energy for refrigeration. By combining systems, solar energy can be used to heat buildings in the winter and cool them in the summer.

The sun also can be used to provide energy that is converted directly to electricity. The technological devices needed for this conversion are called *photovoltaic cells.* Simply put, the sun's

**Natural resources** hold great potential for the development of energy. Evidence of the heat within the earth can be seen in the eruptions of Old Faithful Geyser, above, left, and a volcano, above, center. Windmills like the one shown above, right, have long been a source of energy for pumping water used in agriculture. Other well-established sources include hydroelectric energy from dams, right, and oil delivered through pipelines, below. (Credits: above, left and right, American Petroleum Institute; above, center, Hawaii Visitors Bureau; right, U.S. Department of the Interior/ Bureau of Reclamation; below, Sun Company, Inc.)

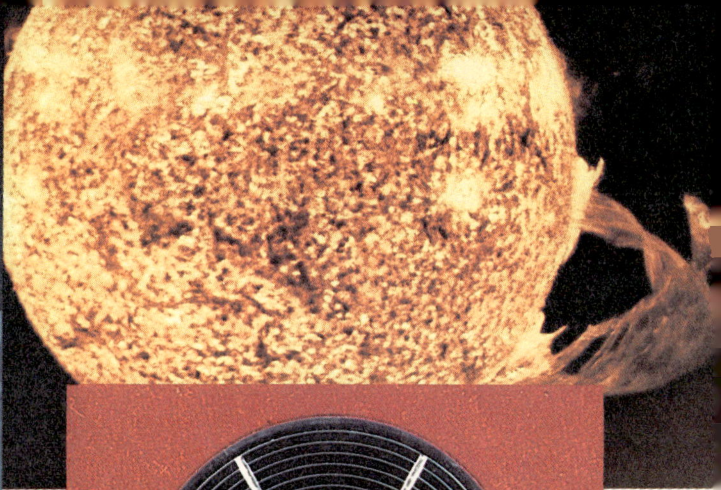

**Atomic and nuclear sources of energy** are already in use and hold some future potential. A major source of energy from radiation is, of course, the sun. In the photo above, left, mirrors are used to reflect the sun's heat into an energy plant. Above, right, is a photovoltaic cell that turns the sun's rays directly into electricity. At left is a photo of plutonium, the material used as a source for atomic energy. (Credits: above, left, Sandia Laboratories; above, right, NASA; center photo, Sandia Laboratories; below, Los Alamos National Laboratory)

**A possible future source of energy** lies in the oceans. Experiments are under way that are aimed at deriving electric power from the waves and tides of oceans. (Courtesy Florida Division of Tourism)

energy acts directly upon material in the cells to start the flow of electricity. Scientists dream of placing photovoltaic cells in space where the sun's energy is much more powerful than on earth. This is because the earth's atmosphere does not filter the sun in space. The power generated from these space ships would be transmitted to earth as microwaves.

Another natural source of energy is the water in the ocean. The level of ocean water near the shore rises and falls in regular cycles. These cycles are known as tides. The force of the water moving in and out of land areas has a great potential for the generation of energy. A number of systems of technology are being developed to harness the power of tides.

Still another potential source of natural energy is the heat of the earth itself. This is known as *geothermal energy*. A number of geothermal systems are already in use. Pipes are driven into the earth. The hot substances found deep in the earth are applied to the pipes. Water pumped into the pipes turns to steam that is used to generate electricity. The same heat source, of course, can also be used to heat buildings. For an example of how geothermal energy is produced, think about geysers, such as Old Faithful in Yellowstone National Park. This is a natural fountain that spews hot water warmed in the earth.

**Fuels.** All of the energy sources described above are natural and "clean." This means that no burning is needed to transform the sources into energy. In addition, because there is no burning, there are no waste products to dispose of in the production of energy. Most energy used by people, however, results from some form of burning, or *combustion*. Combustion uses fuels as the source of energy.

Historically, the main fuel used to produce heat was wood. The burning of wood has, literally, consumed many forests around the world. During this century, wood has been replaced by other fuels, including coal, oil, and gas. In addition, nuclear energy is now used to generate electricity. Today, nuclear energy is a source of more than 10 percent of the electricity used in the United States.

Generating energy from fuels like those described above produces major benefits. However, energy generation also has its problems. The most dramatic of these problems include pollution and waste. For many years, coal was the main source of

**Technology searches continuously** for new sources of energy. The facility in this photo produces gas from coal. (Courtesy American Petroleum Institute)

energy generation. However, burning coal generated smoke and caused many other problems. The area around a generating plant became dirty from the settling coal dust, or soot. The air containing the coal soot was unhealthy to breathe. The problems with coal led to the conversion of many generating plants to oil and gas as sources of energy. These fuels, however, also caused problems. Burning oil and gas discharged unhealthy chemicals into the air. Also, as discussed below, oil and gas as fuels are threatened by shortages and rising costs.

Coal and nuclear fuels present a different problem: Waste is left over after the fuels have been consumed. The ashes left over when coal is burned occupy a lot of space. They are also dirty and contain dangerous by-products. Disposing of coal waste is a long-standing problem in urban areas. Nuclear waste presents problems that have not yet been solved. The equipment and fuels used in a nuclear power plant give off dangerous radiation. Radiation is a form of energy that can cause diseases in the human body. The problem with nuclear waste is that it remains active for more than 100,000 years. It must be stored in ways that contain and control the radiation. Scientists are still looking for a technology to handle nuclear waste satisfactorily.

A potential energy source that many scientists hope will have fewer problems is hydrogen. Hydrogen is a lightweight gas that becomes a liquid when stored under heavy pressure. As liquid hydrogen is released, it becomes a gas again. The gas can be burned as a fuel. The advantage of such a fuel is that it is clean and safe. Burning hydrogen simply forms steam, which is discharged into the air as part of the climate system within an environment.

## Supplies of Resources

For some resources—both materials and sources of energy—supplies could become a problem. Shortages are expected in the future because the resources involved are *exhaustible.* Exhaustible means that there is only so much of an item available and that, in time, the supplies will run out.

Exhaustible resources include those that come from nature. For some resources, nature only produced a certain amount of materials. For example, oil supplies from all over the world are limited. This is also true for coal and natural gas. For well over 100 years, people have used these exhaustible resources without concern for total supplies. Technologies have mined iron ore and coal to make steel. Coal has been used for heat and for power. Oil has been used for power and as a basis for plastic and other manufactured items.

**Wood for home construction** is an example of a renewable resource. (Courtesy American Petroleum Institute)

So far, supplies have met demands. However, existing supplies will not last forever. People concerned about the future recognize that certain key resources are exhaustible. Work is under way to find other, alternate sources of raw materials and energy. The potential use of hydrogen to replace gasoline is a good example of these efforts.

Some resources are *renewable.* Renewable means that the items can be replenished in some way. Most agricultural resources are renewable. People eat crops such as corn and wheat every year. Every year the farmers plant new crops. These food resources, thus, are renewable.

Although the cycle of renewal is slower, wood is also a renewable resource. As trees are cut and processed into lumber or paper, new trees are planted. A tree doesn't replace itself every year. However, large areas are being managed so that new trees are planted and cut every 15 to 25 years.

A final category of resources is made up of *inexhaustible* materials or sources of energy. Examples include the wind, running water, the sun, the tides, and geothermal heat. As long as the earth supports life, these sources will remain available. Clearly, people need to use these resources as much as possible to replace exhaustible supplies. Strong efforts are being made to convert from exhaustible materials and sources to those that are inexhaustible or renewable. Altogether, these efforts are known as *conservation* programs.

**Solar power** is considered to be an important potential source of energy. This facility is designed to reflect the sun's rays to produce usable energy. (Courtesy Sandia Laboratories)

## Supply Problems

Having resources available can be one thing, getting them to the point where they are needed can be something entirely different. For example, scientists knew, for many years, that vast deposits of oil existed on the northern slope of Alaska. These deposits lay under thick, permanent layers of ice. As long as there was a plentiful supply of cheap oil available, people reasoned that drilling oil from the North Slope would be too expensive. However, following the shortages and price increases of 1973, oil companies thought more about these resources. Billions of dollars were spent to build a pipeline that runs the entire length of Alaska, more than 800 miles. The oil has to be heated to keep it from sticking in the pipeline. However, technology has found a way to get oil from the North Slope to a port in southern Alaska. From there, tankers carry the oil to refineries in the "lower 48" states.

**Recycling can play an important role** in the conservation of resources. The facility in this photo recycles aluminum and reuses the basic metal. (Courtesy Aluminum Company of America)

A partial answer to shortages and increasing costs has come through *recycling*. Recycling is the reuse of materials that have already been used. Examples include aluminum cans and paper. Most paper is made from trees. However, certain types of paper can be made from paper that has been used. Aluminum cans can be melted and reused. Major recycling programs exist for many types of metals, including iron, steel, copper, and brass. Glass is also being recycled.

Another way in which resources are conserved is through reuse of containers. A number of states require that bottles be returned to the stores where they were purchased. These materials are then reused. Reusing containers also helps to reduce the amount of waste generated in populated areas.

## SYSTEMS OF TECHNOLOGY

To serve people, technologies are grouped into systems. In this sense, a *system* is a combination of elements. A system of technology involves a combination of techniques and resources to produce a specific result. This result, usually, is different from the results that can be delivered by any other technologies.

In the modern world, systems of technologies become foundations for change and progress. To illustrate, consider the transportation system that you and other people throughout the country use regularly. Transportation is a vast combination of many technologies. To achieve land transportation, people must build roads and bridges. Also required are railroads, trains, and railroad stations. Airborne transportation requires aircraft, airports, and control systems to guide the planes. The transportation system also takes in all of the complex technologies used to build and sell cars and trucks. The entire petroleum industry supplies fuels and oils used in transportation.

Any given system, then, can use many technologies. In Chapter 4, you will be introduced to the technology of mechanics. This technology has been applied again and again to many different systems of technology. Understanding the multiple uses of a technology will help you as you prepare for a career.

## Summary

Technology depends on both human and natural resources. Human resources include the roles of people in developing the products of technology: demand, invention, implementation, and control. Natural resources are materials or energy supplies found within the Earth's natural environment.

Demand is an essential part of technology. People must need or want some product of technology before the other roles—invention, implementation, and control—can take place. Demand is closely related to the cost and availability of resources and end products.

Invention is the mental process for putting ideas to work. People dream of a better way to do something and apply their intelligence to make the idea a reality.

Implementation includes all of the steps that make technology available. Activities range from training people to manufacture products to distributing the end products to users.

Control is the way a society distributes and uses the products of technology. Because demand for technology usually exceeds supplies, controls are necessary. Controls may be applied in the marketing of a product, or a society may pass laws to control the use of a technology. Controls are a part of the resources and products of any technology.

Natural resources fall into two categories: materials or sources of energy. Materials are all substances used in products of technology—solid, liquid, or gas. Sources of energy are forces or potential forces used to produce energy. Materials and energy are related closely as resources of technology.

Raw materials are basic substances used to create the products of technology. Clay and lime, for example, are raw materials for bricks. Soil, sun, and water are raw materials in the technology of agriculture.

A few sources of energy can be used in their natural state. These include wind, running water, the sun, or the heat of the earth. Other sources, such as fuels, must be transformed to produce energy. Combustion, or burning, is the most common way of transforming a fuel to energy-producing heat.

Supplies of resources are an important factor to any technology. Some resources are exhaustible. This means that a limited quantity of that resource is available. Renewable resources are those that can be replenished. An example of a renewable resource would be wood: Trees can be planted and grown to produce more wood. Inexhaustible resources are those materials or energy sources that will exist as long as the earth supports life. A major challenge of modern technology is to manage the supplies of resources to avoid shortages.

Changes in modern life are seldom the result of a single technology but rather a system of technology. Technologies share knowledge, techniques, inventions, and even resources to produce results.

## Technology Terms

natural resource
human resource
demand
petroleum product
crude oil
monetary system
medium of exchange
economy
economic system
invention

## Using Technology

1. In undeveloped parts of the world, there are still people who do not use written language to communicate. All important knowledge is passed on orally, or is spoken, to group members. In these societies, would computers be useful? Discuss your answer in a short composition.

2. It takes 17 trees to produce one ton of newsprint (an inexpensive paper used for newspapers). Suppose your school began a recycling program and that three tons of newspaper were collected weekly. How many trees could be saved in a month with this program? If you were asked to organize this recycling program, how would you begin? What businesses would you contact? How would you collect the old newspapers? Discuss your plan in a short composition.

3. Geothermal energy is a technology that uses the natural heat of the Earth. Look in your library's reference books for more information about how geothermal energy is produced. Draw a schema (diagram) showing how geothermal energy is captured and used.

implementation
refinery
surplus
marketing
legislation
regulate
zoning law
environment
pollution
environmental protection
material
source of energy
raw material
solar energy
photovoltaic cell
geothermal energy
combustion
exhaustible
renewable
inexhaustible
conservation
recycle
system

# 11

# PHYSICAL TECHNOLOGY

Technology undoubtedly started with basic, physical improvements in the way people supported themselves. In earliest times, the first needs of people were for tools to help them survive. People had to protect themselves against natural enemies and against their surroundings. In stages, people developed tools for hunting, for warming themselves, and for providing shelter. Early tools, in turn, were used to develop advanced tools.

The four chapters in this part of your text review the basic principles of physical technology. These principles provide a basis for today's life-supporting technologies.

Chapter 4 deals with the technologies of mechanics. Some of the most basic tools and labor-saving devices are described. Included are the hammer, the wheel, the lever, the pulley, and gears.

In Chapter 5, you carry forward into the technologies of power. The chapter deals with most basic sources of power for modern society. These are the external combustion engine and the internal combustion engine.

Chapter 6 deals with the energy that helps drive modern civilization. Stress is placed upon the technologies that produce electricity.

Chapter 7 traces the evolution of structural technologies to meet people's needs to house themselves. Also covered are shelters in which people work and get together for entertainment.

# 4 TOOLS AND MECHANISMS

## Your Learning Job

After you have read this chapter and completed the work assignments, you should be able to:

- ☐ Define the term tool and explain how tools have helped people.
- ☐ Define mechanisms and explain how mechanisms are different from tools.
- ☐ State what is included in the technology of mechanics.
- ☐ Identify and explain the key principles of mechanics.
- ☐ List several simple tools and explain how they are used.
- ☐ List several mechanical devices and explain how they work.

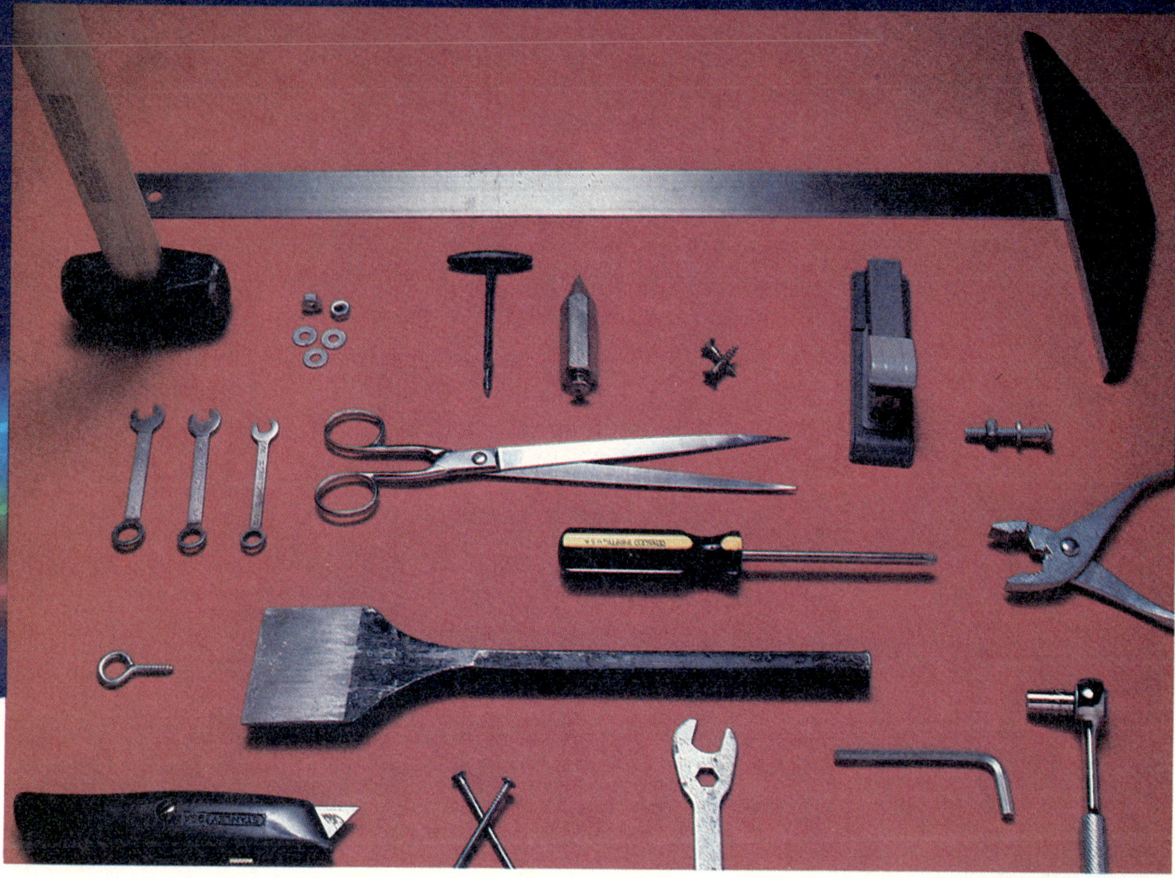

## PEOPLE AND TOOLS

**Tools increase the power of people** to produce work. Tools have been designed and improved continuously since prehistoric times.

The use of tools marks a major difference between the ways of people and those of animals. Animals are limited in what they can do by their basic physical capabilities. Any given animal can only run so fast, carry so much, or travel so far in a given day. People, on the other hand, have been able to develop and use tools to expand their natural capabilities.

With tools, people can shape objects and build structures that they could not produce with their natural physical capabilities alone. With bare hands, people couldn't build furniture. They couldn't construct homes. Neither could they create highways on which they could travel hundreds of miles in a single day. Certainly, people could not fly on their own.

People have expanded their capabilities by developing *tools* and *mechanisms*. In this sense, a tool is any object that is used directly by people to produce specific products or results.

Examples of tools include hammers, saws, wrenches, and shoehorns.

A mechanism performs a slightly different function. Mechanisms are devices that people use to produce power or energy. These power and energy outputs are beyond basic human capabilities. Examples of basic mechanisms developed and used by people include levers, wheels, pulleys, and gears.

Both tools and mechanisms can be grouped under the technology of *mechanics*. Mechanics involves the use of physical, or mechanical, devices.

**Devices for digging and lifting** are among the many power tools that apply the principles of mechanics, as illustrated in this photo showing pipeline construction. (Courtesy Sun Company, Inc.)

## Mechanical Principles

The technologies of mechanics resulted from human discovery. Early people observed and experimented to develop the kinds of tools that we take for granted today. The principles behind mechanics are relatively simple. You can observe these principles wherever people work, apply power, or use energy. Knowing these principles will improve your ability to use tools and mechanisms, now and in future work situations. Some of the key principles of mechanics include:

- It is easier to roll weight than to lift or carry heavy objects. Almost every form of technology has benefited from the discovery of this principle. The principle, in turn, led to the invention of the wheel.

- A physical object that is moving delivers more force than one that is still. A stone lying on the ground delivers the force of its own weight. But, if you pick up that same stone and swing it, you develop force. The stone can be used to drive a nail or for other work. As an experiment, you can drive a nail partially into a piece of wood. Take a stone or the head of a hammer. Place this object against the head of a nail, permitting it to apply its own weight to the nail. Very little, if anything, will happen. Now swing the hammer or stone and hit the nail. Considerably more force is exerted when the object is moving. This principle is known as *momentum*. The use of momentum is behind the design of many tools and mechanisms.

- Falling objects gain momentum. Anything that is loose in the air will fall to earth. This is because the Earth exerts a force

**The wheel has rolled on** through the years as people have depended more and more on personal transportation. Modern highway systems demonstrate how an entire civilization has been built around personal transportation, in the form of the private automobile. (Photo by W.R. Grace and Co.)

upon loose or falling objects. This force is known as *gravity*. The further an object falls, the more momentum it gathers as it is drawn by gravity. This principle applies in the design of tools or mechanisms: It is easier to push down on an object than it is to lift up.

- *Leverage* increases the power delivered from the use of force. Leverage is accomplished through use of a *lever.* A lever is a rigid (stiff) object that is usually long and narrow. One edge of the lever is placed at the point where force is needed. Force is applied to the other end. The presence of the lever increases greatly the amount of the force. You use leverage when you turn a handle to cause movement of some object. Examples include doorknobs and the handles of faucets.

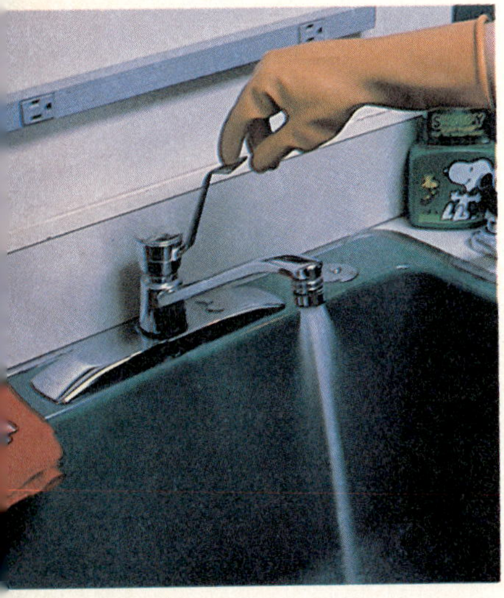

**The principle of leverage** is applied in many devices you use in your everyday life. You might say leverage is as common as the kitchen sink.

Knowing these simple mechanical principles will improve your ability to understand and use tools and mechanisms.

## TOOLS

The development of tools appears to have followed people's ideas about the structures they needed. Safety has also been a factor. The principles used in developing tools have undoubtedly been related to experiences in hunting and in processing food.

Through the centuries, people have developed many, many tools for many, many purposes. It would be impossible to review all of these. However, looking at a few simple, popular tools will illustrate the technology of mechanics. These include:

- The hammer
- Cutting tools
- Grinding tools
- Drilling and milling tools
- Fasteners.

### Hammer

The *hammer* uses combined principles of technology in a simple, direct way. The principles are momentum and leverage.

A hammer, as you undoubtedly know, has a metal, usually steel, head attached to a handle. There are many types and styles of hammers. Some hammers are designed to shape metal. These are used by craftsmen for such jobs as straightening dents in automotive fenders. These hammers, which have wide, flat heads, are also used to make decorative metal objects. Possibly the most widely used type of hammer is known as the *claw hammer.* This is designed for driving and pulling nails. There are also *sledge hammers.* These have heavy heads that develop extremely powerful momentum. Sledge hammers are used for such jobs as driving stakes or breaking up pavement.

A variation of the same principle used in the hammer is the axe. This is a combination of hammer-like force and a cutting tool. Cutting tools are described in the section that follows.

When you use a hammer, you apply momentum by moving the head in the direction where you want it to strike. The direction of movement for the hammerhead is in the form of an arc formed by the length of the handle. The length of your arm may also add to the length of the hammer's swing. This can be important because, as the hammer swings, leverage is added to momentum. The greater the distance from the point at which the swing originates to the hammerhead, the greater the leverage.

Leverage is also used when you pull nails with a claw hammer. You can try this. Take a nail that has been driven most of the way into a piece of wood. Try to pull the nail with your fingers. The nail will probably not move. Now place the nail between the claws of the hammerhead and pull back on the handle. Removing the nail becomes relatively easy. You have made good use of leverage.

**Momentum and leverage are at work** when a hammer is used. In this photo, the hammer is being used to drive a chisel, a type of cutting tool.

## Cutting Tools

Building materials come in a variety of sizes and shapes. It is often necessary to cut raw materials so that they fit the structure on which they will be used. Cutting tools are needed. Cutting tools were also needed as early weapons. As people figured out ways to protect themselves against animals and to hunt, some principles became clear. One is that a sharp object does a better job of penetrating a surface (cutting) than a dull one. Undoubtedly, early hunters put this principle to work by sharpening sticks to make spears. It is known that early humans used natural stone to make arrowheads and the first knives. The stones used were materials that could be chipped or broken to form sharp surfaces. Many early cutting tools have been found that were made of a stone called flint. Arrowheads of flint could pierce animal bodies. Knives could be used to skin animals to produce hides for shelters and clothing.

As people learned to work with metals, particularly iron and steel, their cutting tools became more sophisticated. Great advances in the design and construction of buildings were possible because of the invention of the *saw.* As you probably know, a saw is a metal cutting tool. A saw has a series of sharp points, or teeth. When a saw is moved over a wood surface, the points keep eating away bits of the surface. Eventually, the saw cuts entirely through the material. When people invented saws, they gained the ability to construct buildings to meet their specific needs.

The knife, of course, is a general-purpose cutting tool. People have come to use knives as essentials in food preparation, eating, and general handiwork. Many people whittle, cutting forms from blocks of wood, as a hobby.

**A cutting tool** commonly used in working with wood is the plane. A cutting blade at the bottom of the plane shaves away a portion of the surface of the wood.

Other woodworking tools have been developed to shape, rather than simply to cut through, wood. For many years, the axe was a major tool in securing wood for buildings. On the American frontier, entire logs cut down with axes were put into place to form log cabins. In cities, logs were trimmed into large beams used in construction. A tool used for this purpose was the *adze*. The adze is an axe-like device, except that the cutting blade is shaped so that it cuts across the direction of the axe-handle swing. The adze was used to trim and shape sections of logs to form beams.

Another important tool for shaping wood is the *plane*. A plane has a flat surface that is moved across a piece of wood. A cutting edge sticks out from beneath the plane, trimming small amounts of wood with each stroke.

After people began using metal for their structures, it was necessary to develop tools to cut metal. Heat is the main technique used in tools designed to cut metal. Special gases or electricity are used to generate the heat. Modern cutting torches can

**Relatively soft materials,** such as paper, are cut with special tools like those shown in this photo.

cut rapidly through even the strongest metal objects. Saws are available to cut metal. But sawing metal is relatively slow and is limited to small jobs.

Softer materials can be cut with a different approach. *Scissors* are used. Materials such as paper or cloth are commonly cut with scissors. Scissors represent a method of cutting in which two metal surfaces are brought together. The pressure that leads to the actual cutting uses leverage.

There are, of course, other devices that cut materials used to produce shelter, furniture, or clothing. The principle, however, remains the same. Materials must be cut and shaped to fit the product or job on which they will be used.

## Grinding Tools

*Grinding tools* also are used to shape wood and metal materials. The principle of grinding is relatively simple: Two materials are used. One of the materials is harder than the other. The surface of the harder material is rubbed against the surface of the softer material. Some of the softer material wears away with each stroke. The stroking, or rubbing, continues until the end product, known as the *workpiece,* reaches the desired size and shape.

Grinding tools offer a good example of relationships among technologies. Grinding techniques are used to sharpen cutting tools. After they have been used for some time, cutting tools such as knives, saws, or scissors become dull. They don't cut as well as they did when they were first made. Grinding tools can be used to put sharp edges on these cutting tools. Examples of grinding tools include sandpaper, files, and grinding wheels.

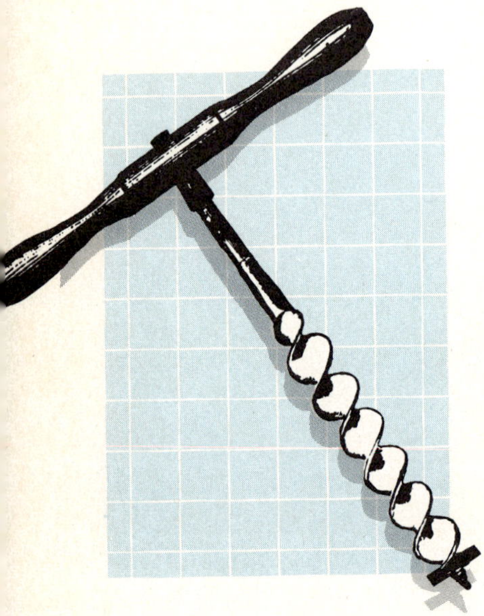

**A basic drilling tool** is the auger. When twisted into a wood surface, the auger drills a hole.

## Drilling and Milling Tools

In some ways, the processes of *drilling* and *milling* combine the work of cutting and grinding. In drilling, a rotating device, called a *bit,* is brought into contact with a workpiece. As the bit turns, it eats away a hole through the workpiece. Thus, the hole that is drilled is cut or ground.

In milling, a bit similar to that of a drill is used. However, the bit can cut to the side as well a directly down. The workpiece is moved while the bit turns. Portions of metal are cut or ground away. The result is that the workpiece is shaped into a desired form through milling. In milling, the workpiece is placed firmly in a special holding device called a *jig*.

Another tool for shaping a wood or metal workpiece is known as the *lathe*. Lathes are used to shape rounded workpieces. A workpiece is fastened into a rotating holder known as a *chuck*. Cutting tools are then placed in contact with the rotating workpiece, taking away unwanted material.

Holes can also be put in relatively soft material with a *punch*. Punching is used commonly for such materials as paper or thin metal. The tools are known as a punch and *die*. The die is a device made of hard metal. It has a hole in the position where the punch will pass through. The material is placed between the punch and die. As the punch is lowered, holes are punched out and the material removed is dropped away. The bits of paper or other material created when holes are punched are called *chad*. For large volumes of punching-type operations, special drill bits, with hollow centers, can be used.

**Alternative methods for drilling holes** include both hand and power tools.

## Fasteners

Fastening means joining. A *fastener* joins two or more workpieces to make them part of a single structure. One of the most basic inventions in physical technology is used for fastening. This is the *screw.* A screw is a narrow, round, pointed metal device that is driven into working material by rotation. The threads of the screw hold the working materials together because the raised portion of the threads overlaps with the actual materials.

The tool used to *drive* a screw into working material is called a *screwdriver.* The screwdriver has a tip that fits into the head of the screw. The bit of the screwdriver is rotated either by hand or through use of power.

The heads of screws and the bits of screwdrivers must have matching shapes. Two basic approaches are used. The head of the screwdriver can have a straight slot. This is simply an indented cut into the head of the screw. A straight-tipped screwdriver is used.

Another type of screw head is known as the Phillips head. Phillips screws have crossed slots in their heads. The screwdrivers also have crossed blades that fit into these slots. The advantage is that a greater amount of pressure can be applied across two slots than across one. Phillips head screws are usually used in mass-produced items. That is, the Phillips heads are designed for use with power-driven screwdrivers.

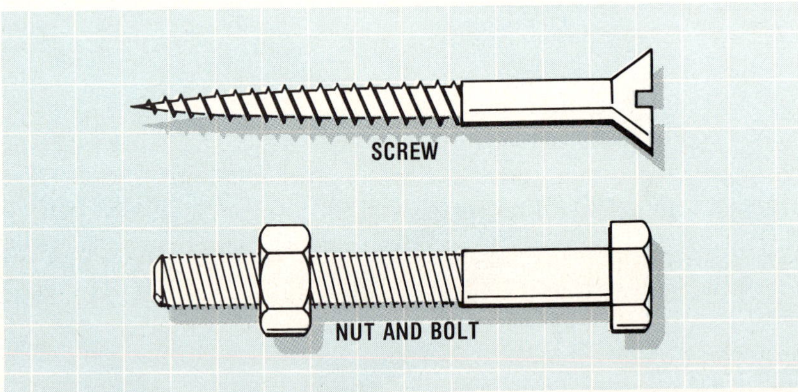


**Fastening devices** include the screw, shown above, right, and the bolt. Each of these threaded fasteners does a different job.

Another threaded fastener is the *bolt*. A bolt resembles a screw. The main difference is that the screw has a pointed tip. The bolt has a flat tip. Bolts may be used in two ways. One way is to *tap* a hole into a metal workpiece. A tap is a tool made from hard metal. The tap creates a threaded hole. The threads in the hole match those of the bolt. When the bolt is tightened into the threaded hole, it holds objects in place. That is, materials between the head of the bolt and the threaded hole are held in place by the bolt.

Under a second approach, a bolt is not driven into working material. Instead a bolt is passed through matching holes into two or more pieces of working material. The bolt is then held in place by a *nut*. A nut is a flat piece of metal with a threaded hole that matches the threading of the bolt. A nut is tightened in place at the end of a bolt to hold workpieces together.

When workpieces to be joined are alongside each other, a separate type of fastener may be needed. This is the *bracket*. A bracket, basically, is a flat piece of metal (wood can also be used) in which holes have been drilled. When screws are driven into the holes of brackets, the workpieces are held together. Brackets are very often used to hang shelves from walls or in construction of cabinets or bookcases. Brackets that hold shelving can be bent. This makes it possible to attach one side of the bracket to a wall and the other to the shelf.

When a workpiece to be joined to a structure has to be able to move, a *hinge* is used. A hinge is a two-part fastener that permits each portion to swing, or rotate. Hinges are used to hold, or hang, doors. Hinges are very important devices for fastening movable parts into structures.

To join metal workpieces, techniques using heat are necessary. One method is known as *soldering.* Soldering is used primarily for small parts, such as electrical or electronic connections. A metallic material that melts easily is placed at the connecting point. Heat is applied to melt the material. When the melted metal cools and becomes a solid material, it holds the connections in place. Solder is usually supplied in the form of a wire. Heat is applied through use of a soldering iron.

*Welding* is done with torches. In effect, welding is the opposite of metal cutting. A rod of metal is melted into a seam between two metal workpieces. When the melted metal becomes solid, the two workpieces are held together.

**Metal workpieces** are joined through a technique known as welding, a skilled craft. (UNRWA)

## MECHANISMS

A mechanism, as indicated earlier, is any device that increases or applies force to perform work. Mechanisms are also called *mechanical devices.* There are so many mechanisms and mechanical devices that entire, large books are devoted to this subject. The discussion that follows deals with basic principles only. Only simple mechanisms are described to illustrate the principles. The mechanisms discussed below are:

- The lever
- The wheel
- The inclined plane
- The pulley
- Gears.

### The Lever

To understand the principle of the lever, think about a seesaw. The typical seesaw consists of a board mounted to a central pipe or beam. The seesaw board is a lever. The point at which the lever rotates is known as a *fulcrum.*

In a seesaw, the idea is that the weight of the people at each end should be balanced. If one person is heavier than the other,

**Technology and child's play** are combined in the seesaw, which provides an everyday example of the use of leverage.

that person will lift the other off the ground. If both people weigh the same, they can ride up and down by pushing up with their legs.

The principle of leverage is based upon the relationship between the lever and the fulcrum. This relationship depends on the position of the fulcrum. If the fulcrum is in the center, power applied and work delivered are equal: The power applied to the lever on one side of the fulcrum equals the work delivered on the opposite side. A seesaw uses the principle of leverage in a balanced situation. That is, the fulcrum is in the middle of the lever. However, if the fulcrum is moved to one side of the lever, the work delivered increases greatly in relation to the effort applied. This increase is in relation to the relative distance of the fulcrum between the ends of the lever.

The illustration above shows a simple example. A long beam of wood is being used to move a large rock. The fulcrum is a second, small rock placed close to the object to be moved. In the example, the fulcrum is at a point equal to one-fourth of the length of the lever. Thus, the power delivered is four times the force applied. Assume, for example, that a 150-pound person leaned full weight on the long end of the lever. The work delivered at the other end of the lever would be equivalent to some 600 pounds.

Leverage also works the other way. When you swing a hammer, for example, the fulcrum point is your elbow or shoulder. Thus, the force exerted by your muscle is multiplied many times.

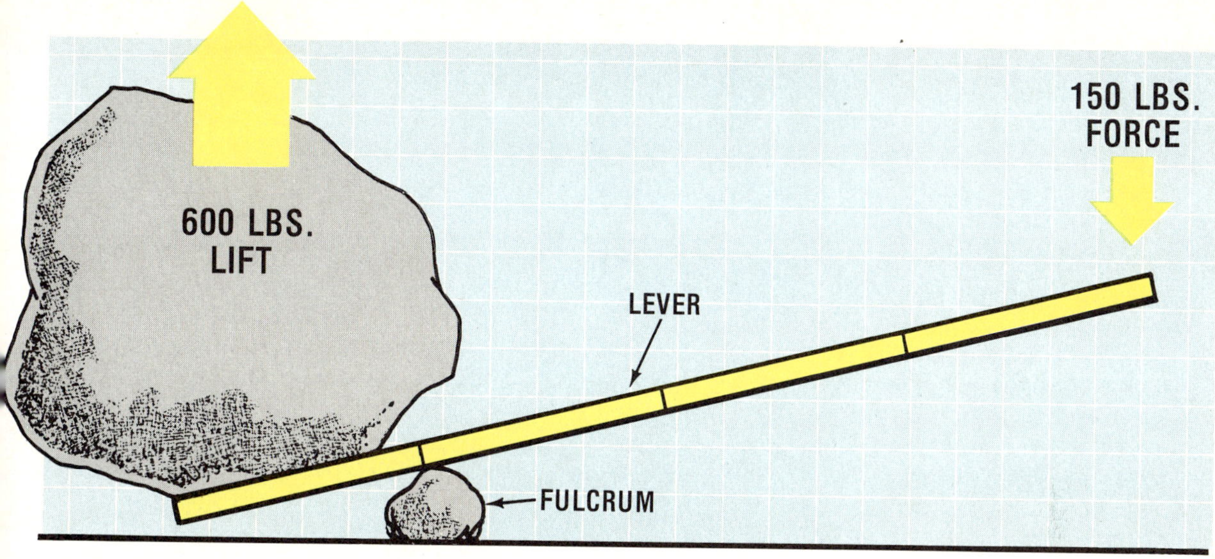


**The lever and fulcrum** help to increase the amount of work performed through the application of force. As shown in this illustration, the position of the lever on the fulcrum serves to multiply the force that is applied.

Leverage is used in many other tools with which you are familiar. These include the claw hammer and scissors. The lever—and the principle of leverage—are applied in many of the power-producing devices described in the chapter that follows.

## The Wheel

The *wheel* is the best known and most effective use of leverage ever devised by people. The wheel needs no description. It is part of your everyday life. A wheel, basically, is a circular structure that rotates around a central point, or *hub*. The wheel is held in place on a device known as an *axle*. The distance between the axle and the outside of the wheel, or *rim*, marks the use of leverage.

The leverage possible through the use of the wheel is multiplied because *friction* is reduced greatly. Consider a large container weighing 1,000 kilograms, or kilos (k). To drag this container along the ground, the energy required would be far greater than the weight of the object. However, suppose the container were placed upon a simple wagon with a flat bed. The bed of the wagon is mounted over four wheels connected by two axles.

Simple carts of this type were invented by people thousands of years ago. The discovery of the wheel is one of the great technologies of all time. A 1,000 kilogram load, for example, may be pulled along a level surface with 100 kilograms of force. In

**Heavy loads** can be pulled with greater ease when they rest upon wheels than when the materials have to be carried by hand. The wheel is one of the oldest inventions developed to make work easier and more productive.

other words, the wheel has the effect of multiplying force applied to produce work. This multiplying factor can be as much as 10 times. The wheel is a basic device that has made modern civilization possible. The wheel is so commonplace that it does not need further explanation. However, you should appreciate the great advantage made possible by the wheel in terms of work produced.

## The Inclined Plane

An *inclined plane* is, basically, an uphill surface. In structures such as buildings, bridges, or roads, inclined planes are sometimes called *ramps.*

An inclined plane, quite simply, makes it easier to move weights to higher levels, as compared with lifting those weights. To illustrate, assume an incline in which the length of the ramp is three times the height. Thus, if the ramp were 30 meters long, the increased elevation would be 10 meters. Assume that the purpose of the ramp is to move materials from one level of a building to another. It takes approximately one half as much effort to pull a load up a ramp as it does to lift it straight up. The inclined-plane principle is used in situations such as the loading of ships, trucks, and airplanes. When ships are tied to docks, inclined planes called *gangplanks* are used. Passengers can walk up and down the gangplanks with far less effort than climbing ladders or stairs. People who carry cargo into ships can also move more easily. The conveyors used to load baggage onto airplanes also follow the inclined-plane principle. The same is true for loading ramps used in placing cargo into trucks.

**The inclined plane** makes it easier to move materials to different levels, as compared with lifting or lowering the same items.

**An inclined plane,** positioned by a system of pulleys, eases the loading and unloading of specialized ships that carry freight from foreign countries. (Courtesy U.S. Customs Service)

## The Pulley

The *pulley* applies the principle of the wheel to the control of weight to be lifted upward (vertically). Pulleys are also used to transfer power in machinery. A pulley is a wheel that has a channel, or cutaway portion, in its rim. This channel guides the movement of a rope or belt that causes the pulley to rotate.

Pulleys are used to develop mechanisms, or systems, for lifting weights vertically. When lengths of rope are threaded through multiple pulleys, the weight of the object to be lifted can be distributed. That is, the weight is spread among the pulleys and over the length of the rope. As a result, leverage is gained.

Pulleys are used in two different ways. A *fixed pulley* is one that is fastened to a holding device or bar. Weights can be lifted through use of rope and one fixed pulley. Suppose, for example, you had a bale of hay to lift into the hayloft of a barn. You could suspend a single pulley from a beam that extends out from the top of the barn. You could then attach a rope to the bale of hay and pull at the other end. Using a single pulley in this way, the lifting force would have to equal the weight of the bale. No leverage is gained. The only advantage is that the direction of force is changed. The force applied would be along the line of the ground. However, the direction of the force would be

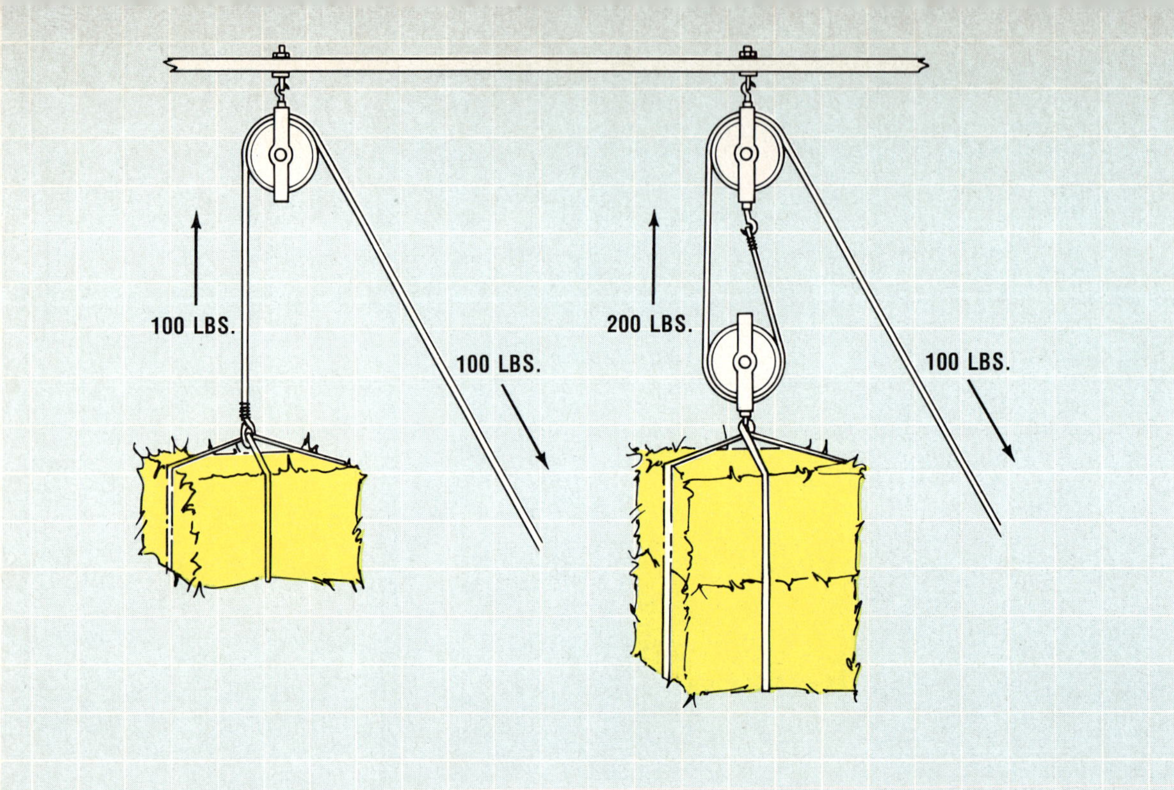


**The number of pulleys used** increases the amount of force applied for lifting.

changed so that the hay is lifted vertically. This operation is shown in the above illustration.

To gain leverage you could add a second pulley. As shown, you could use one fixed pulley and one *movable pulley.* This adds to the length of the rope and spreads the weight of the load. In this system, there is still one fixed pulley. But a second, movable pulley has been added. With one fixed pulley and one movable pulley, the amount of force applied in lifting weight is doubled.

Additional pulleys could be added to the system. Other combinations could include two fixed pulleys and one movable pulley. This arrangement, using three pulleys, would triple leverage. If two fixed and two movable pulleys are used, lifting power would be equal to four times the applied force.

Pulleys are also used in machinery to transfer power. In such situations, the pulley that is connected to an engine or other power source is called the *driving pulley.* Other pulleys that may be linked by the same belt are called *driven pulleys.* The relative size of the pulleys determines the operating speed and power delivered. In this respect, systems using pulleys and belts are similar to those using gears.

## Gears

A *gear,* basically, is a round, wheel-shaped device that rotates on a central shaft. On the outer rim of the gear are a series of ridges known as *teeth.* The teeth are used to enable one gear to mesh, or work together with, other gears or devices. For an everyday example of gears, think of a bicycle. The gears on bicycles are called *sprockets.* The teeth of the sprockets engage in a chain. A bicycle has at least one front sprocket, attached to the pedals, and one rear sprocket, attached to the rear wheel.

When you turn the pedals of the bicycle you are applying *torque.* Torque, basically, is turning force. On a bicycle, the torque is applied from the front sprocket to the rear sprocket, as you turn the pedals.

Notice that the front sprocket of the bicycle is a lot larger than the rear one. This makes it possible for you to apply leverage to cause movement of the bicycle. For each time you turn the large, front sprocket, the rear one may turn three to five times. This increased number of turns for the driven sprocket, as compared with the driving sprocket, produces leverage, or high torque.

The same principle applies when gears are meshed so that their teeth interlock as they turn. The illustrations below demonstrate this principle. The amount of torque delivered is described as a *ratio.* A ratio is a set of related numbers. A driving gear, for example, might have three times as many teeth as a driven gear. The ratio would be noted as 1:3. In reading this, you would say "one-to-three ratio."

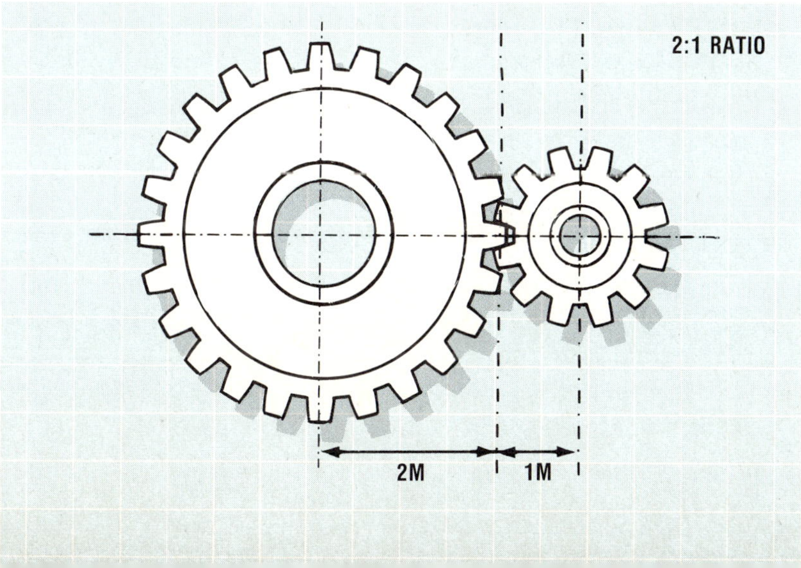


**Gears apply a special leverage** known as torque, or turning force. Torque increases or decreases in relation to the size of the gears.

Tools and mechanisms were, undoubtedly, the beginnings of modern technology. People developed tools and mechanisms to increase the work they could deliver from their own strength. As experience with technology led to a growth of knowledge, people expanded the value of tools and mechanisms by adding sources of power and energy. These principles for adding power and energy are described in the chapters that follow.

## Technology Terms

tool
mechanism
mechanics
momentum
gravity
leverage
lever
hammer
claw hammer
sledge hammer
adze
plane
scissors
grinding tool
workpiece
drilling
milling
bit
jig
lathe
chuck

## Summary

People have expanded their physical capabilities through tools and mechanisms. A tool is any object that can produce a specific result. A mechanism is a device that will increase the power or energy that is applied. Both tools and mechanisms make up the technology of mechanics.

People discovered several key principles of mechanics and applied these principles in making tools and mechanisms. One important observation is that it is easier to roll a heavy object than to carry or lift it. Another discovery was that a moving object delivers more force than a stationary object. Closely related to that idea is that falling objects gain momentum because of the Earth's gravity. Yet another important principle is that leverage increases the power delivered from the use of force.

These principles are applied whenever simple tools are used to perform work. The hammer, for instance, applies both the principles of momentum and leverage. Other simple tools include cutting tools, grinding tools, drilling tools, and fasteners. While many sophisticated tools have been developed to perform complex jobs, the basic principles remain the same.

Mechanisms, or mechanical devices, also apply basic principles of the technology of mechanics.

The lever, like the hammer, applies the principle of leverage. Other simple mechanisms are the wheel, inclined plane, pulley, and gears. The majority of mechanisms, even those that appear complex, use these basic mechanical devices to perform work.

Tools and mechanisms were the beginning of modern technology. The technology of mechanics has allowed people to perform more work than through either animal or human strength alone. The technology of mechanics is a part of most systems of technology. Therefore, understanding the principles of basic tools and mechanisms will help you to understand other technologies.

## Using Technology

1. Identify the principles of mechanics that explain how the following common items work: Spatula, dolly, slingshot, and grandfather clock.

2. Suppose that a driving gear has 24 teeth and the driven gear has 6 teeth. What is the gear ratio?

3. Suppose you and a friend are given the job of loading heavy boxes onto a high platform. What mechanisms might help you do this work? Draw a simple diagram of one such mechanism.

punch
die
chad
fastener
screw
drive
screwdriver
bolt
tap
nut
bracket
hinge
soldering
welding
mechanical device
fulcrum
wheel
hub
axle
rim
friction
inclined plane
ramp
gangplank
pulley
fixed pulley
movable pulley
driving pulley
driven pulley
gear
teeth
sprocket
torque
ratio

# 5 POWER

## Your Learning Job

After you have read this chapter and completed the work assignments, you should be able to:

- ☐ Define the terms power, force, and work as they apply to technology.
- ☐ Outline the steps of a power cycle whose source of energy is gas.
- ☐ Define external combustion and describe how a steam engine works.
- ☐ Define internal combustion.
- ☐ Define combustion and name three conditions that must exist for combustion to occur.
- ☐ Describe the functions of a four-stroke cycle engine.
- ☐ Describe the combustion principle used in diesel engines.
- ☐ Discuss the problems of using combustion engines.
- ☐ List several systems that must work together in an automobile.
- ☐ Describe the difference between turbine and rocket engines.

## THE NATURE OF POWER

**The meaning of power** in a sport like racing can refer to speed and performance. (Courtesy Valvoline Oil Company)

In technology, the words you hear and use have special meanings. It is important to understand the meanings and to use the words accordingly. The word *power* offers a good example. Power can have different meanings, depending on the way the word is used. There can be political power, the power to convince people, or the power of physical strength. In technology, however, power has a specific, limited meaning.

As defined earlier, power is the rate of doing work. The word *rate* indicates that power can be measured. The rate of work describes results, or the amount of work that is done in a given time. So, power is a measure of the amount of work done in a certain amount of time.

*Work* itself has a special technological meaning. In power technology, work is the use of *force* to move something. So, to understand the meaning of work and power in technology, you must now learn about force. Force is applied energy or effort—energy or effort that has been put to use.

**Power for space travel** represents a special, modern use of this technology. Launching of space vehicles depends upon thrust delivered by rocket engines. (Florida Division of Tourism)

You can apply force without getting any work in return. For example, suppose you push as hard as you can on a car that won't start. The car doesn't move. It may be that the brakes are on. Or maybe the car is facing up a hill and you can't make it move. The reason doesn't matter. The point is that, if you don't have movement, no work has resulted. You may feel tired from your efforts. But in the sense of power technology, there is no work without both force and movement.

Now, if the stalled car were facing down a hill, the result could be different. If the brake were off, a light push could deliver big results. Force and movement together equal work.

The greater the movement, the more work is achieved. This means that work, like power, can be measured.

Power technology, then, is made up of a series of parts. As discussed earlier, most technologies are made up of a series of parts. Basically, technology is the practical use of science, or scientific principles. By combining principles of science, technologies take on an identity of their own. The parts of power technology include work, force, and movement. Each of these parts, in turn, has other parts. Force, for example, is applied energy or effort.

Systems of technology, then, result from the way in which scientific and technological parts are combined. Within the broad field of power technology, for example, there are a number of different combinations, or systems, of power.

## SYSTEMS OF POWER

A *power system,* or type of power, applies one or more sources of energy to perform work. Some sources of energy have already been covered. These include wind, water, and heated gases. Other sources and types of energy are described in Chapter 6.

Heated gas is the main source of energy for the power systems in general use today. Any system that uses heated gas as a source for power has certain parts and events. These include:

- Fuel is burned to produce heat. The burning of fuel is part of a process called *combustion.* One result of combustion is heat.
- Application of heat causes gases to expand. In some systems, gas is produced when heated water boils. The boiling water turns to steam. Steam expands so rapidly that it provides force.
- The expanding gas causes movement of some object. This means that work is being done. Power results.
- Power must be converted to some useful force. To have a system of power, you need some planned results. These results, for example, can be the turning of wheels to make a car move.

These steps mark a *power cycle.* The cycle begins with the use of energy to create force. Then force is applied to produce work.

The work produced is a measure of the power that is created. This system of power, remember, is based on the use of combustion to create a force. Two general types of combustion systems are used in energy production:

- External combustion
- Internal combustion.

## EXTERNAL COMBUSTION SYSTEMS

*External combustion* means that heat is produced outside the power device (engine). The most common type of external combustion unit is the *steam engine.* In the operation of a steam engine, a fire applies heat to a *boiler,* a metal tank that holds water. Steam from the water expands into a narrow, tube-like enclosure called a *cylinder.* A cylinder is a place within any engine where energy is changed into force. In the steam engine, the energy of the expanding gas pushes against a *piston.* A piston is a circular device that moves within a cylinder. The piston has solid surfaces that are pushed away from the expanding gas. The force from applied energy causes the piston to move along the length of the cylinder. After the piston moves the length of the cylinder in one direction, steam is fed from the other end. This causes the piston to move in the opposite direction.

As the piston moves, force is transferred through a series of parts. On a steam engine, force is transferred from the piston to a *piston rod.* This is a steel rod connected to the piston. The piston rod, in turn, is linked to a connecting rod. Force is transferred from the piston rod to the connecting rod. In the case of a locomotive, the connecting rod turns a wheel to which it is linked. As the wheel turns, the train moves.

Think back, now, to the definition of work. To deliver work, there must be force and movement. In the case of an engine, the critical movement is accomplished by the piston within the cylinder. This is the movement that counts. The result of this movement may be a rolling of the locomotive along the tracks. But movement of the locomotive is a result of the work within the engine. Movement of the train is not the work that counts—at least not in a technological sense. Movement is a direct result of the application of force. Within a steam engine, force moves the piston.

In the steam engine description so far, steam drives the piston along the length of the cylinder. At this point, the piston has to move in the other direction along the length of the cylinder. The movement of a piston in an engine cylinder is known as a *stroke*. Following the initial stroke, something must be done to move the piston in the opposite direction. In a steam engine, this is done by feeding steam into the opposite end of the cylinder. This causes a second stroke that drives the piston back to its original position. Again, force is transmitted through the piston and connecting rods. When applied to the wheels of the train, the returning rods complete the rotation. This adds to the movement of the wheels that rest on the rails. As the engine goes through repeated strokes, the locomotive moves farther—and eventually picks up speed.

**The external combustion engine** was the major source of power for manufacturing and transportation for many years. External combustion is used to drive the steam engine of this classic railroad train.
(© Allan Seeger/Uniphoto)

The example discussed so far involves the use of steam power to drive a railroad locomotive. Steam engines also played a major role in providing power for production equipment in factories. Power devices that stay in one place and drive equipment are called *stationary* engines. Stationary means that something is fixed; it doesn't move. The workings of this engine are pretty much the same as those for the locomotive. The main difference is in the point to which the connecting rod is attached. Instead of being hooked to the wheel of a locomotive, the connecting rod of a stationary engine moves a *flywheel.* The flywheel, in turn, becomes a source of power for production machinery. This power is transferred to the factory machines through a drive belt. The belt is turned by the flywheel and, in turn, delivers power to a pulley that turns a machine. The machines powered in this way are said to be *belt driven.*

The steam engine is now thought of as an early power source that is not very efficient. Internal combustion engines, described below, have pretty well replaced external combustion units. However, for two centuries, the steam engine was the main source of power for transportation and production. The steam engine made sea travel possible without relying on the wind. Railroads used steam locomotives. Factories depended on steam to run machines. In short, the steam engine powered the Industrial Revolution.

Steam-powered locomotives were a major force in the expansion of the United States across the entire continent.

The steam engine also represented a victory for ideas. With steam power, people no longer had to depend on water or wind. People used their abilities to produce and control power. The steam engine also provided an opportunity to learn about the use of combustion to develop power. As people began to realize the potential of combustion, they began to think of other ways to apply power to make life easier. The steam engine, in this way, created a demand for power. People were ready for the improvements made possible by internal combustion engines.

## INTERNAL COMBUSTION SYSTEMS

*Internal combustion engines* also deliver power by using expanding gas as a source of force. The expanding gas causes movement of a piston along the length of a cylinder. The movement is transferred through a series of rods to deliver power. In short, the principles of expanding gas as a source of energy are basically the same.

The major difference lies in the approach to combustion. The type of gas that expands to provide a source of energy is also different. In an internal combustion engine, burning takes place within the cylinder. The gas that expands to drive the piston is air. The drawings on page 102 illustrate the basic operation of an internal combustion engine. Refer to these as you read the following explanation.

## Combustion

Remember the basic principle: Combustion itself (rather than steam provided from boiling water) is the source of force. Because of this, you should understand the basic nature of combustion. This will provide a basis for understanding how an internal combustion engine delivers power. To have combustion, three conditions must be met:

- There must be fuel.
- Air (oxygen) is needed.
- A required temperature must exist.

Remember these basics of combustion: *fuel, oxygen,* and *temperature.* Fuel is a material that can burn. Fuels are selected to meet the needs of the power system in which they will be used. For steam locomotives, popular fuels were wood and coal. Fuel was burned in a firebox under the boiler. Internal combustion engines use petroleum (oil) products as fuel. The most common fuel for internal combustion engines is gasoline.

Oxygen is one of the gases present in air. Oxygen is essential to life. Oxygen is also one of the necessities of combustion. Nothing will burn unless there is a supply of oxygen present.

The proper mixture of fuel and oxygen also requires a certain temperature before combustion takes place. Each material has its own combustion temperature. These temperatures are different for paper, wood, gasoline, and other fuels. These differences can be valuable in starting and controlling combustion. For example, to build a fire, you may light paper first and use these flames to start the wood burning. This makes sense because paper burns at a lower temperature than wood. Thus, it is easier to start a fire by *igniting* paper first. To ignite means to set something on fire.

The need for temperature in supporting combustion holds the key to controlling and fighting most fires. A common way to put out a fire is to throw (or spray) water on it. The water's temperature is below the combustion temperature of many materials. Thus, water takes away one of the requirements for combustion—temperature. Fires with some types of fuel, such as gas or gasoline, cannot be put out effectively with water. On

**The ingredients for combustion** include fuel, temperature, and oxygen. This photo shows combustion of crude oil trapped within oil shale rock. The oil provides the fuel. (Sun Company, Inc./American Petroleum Institute)

such flames, foam or gases without oxygen often are used to put out fires. These materials take away the supply of oxygen, eliminating one of the requirements for combustion.

The technology of power through internal combustion engines relies on these principles of combustion. Air and gasoline are forced into the cylinder of an internal combustion engine. This mixture is designed for rapid burning. Burning takes place so fast that a small amount of fuel actually explodes. The final need for combustion is temperature. This is created by introducing an electrical spark into the cylinder immediately after entry of the fuel and air. The spark creates the needed temperature.

The great heat that is produced by the burning causes the gas in the cylinder to expand rapidly. This drives a piston, producing force to turn the engine. The power of the turning engine, in turn, is transferred to the wheels of the auto.

## The Four-Stroke Cycle

The burning of fuel within an engine is more complex than heating water. Therefore, the typical gasoline engine requires multiple *cycles,* or up and down movements of pistons, to deliver power. The most common combustion system for gasoline engines is the one illustrated. This uses a *four-stroke cycle.* For clarity, separate illustrations show what happens during each piston stroke in the combustion cycle of a gasoline engine. The strokes within the typical gasoline engine operating cycle are:

- Intake
- Compression
- Power
- Exhaust.

**Intake stroke.** During the *intake stroke,* the piston moves downward. This creates a suction that draws in a mixture of gasoline vapor and air. The entire space within the cylinder fills with a gas mixture of fuel and air.

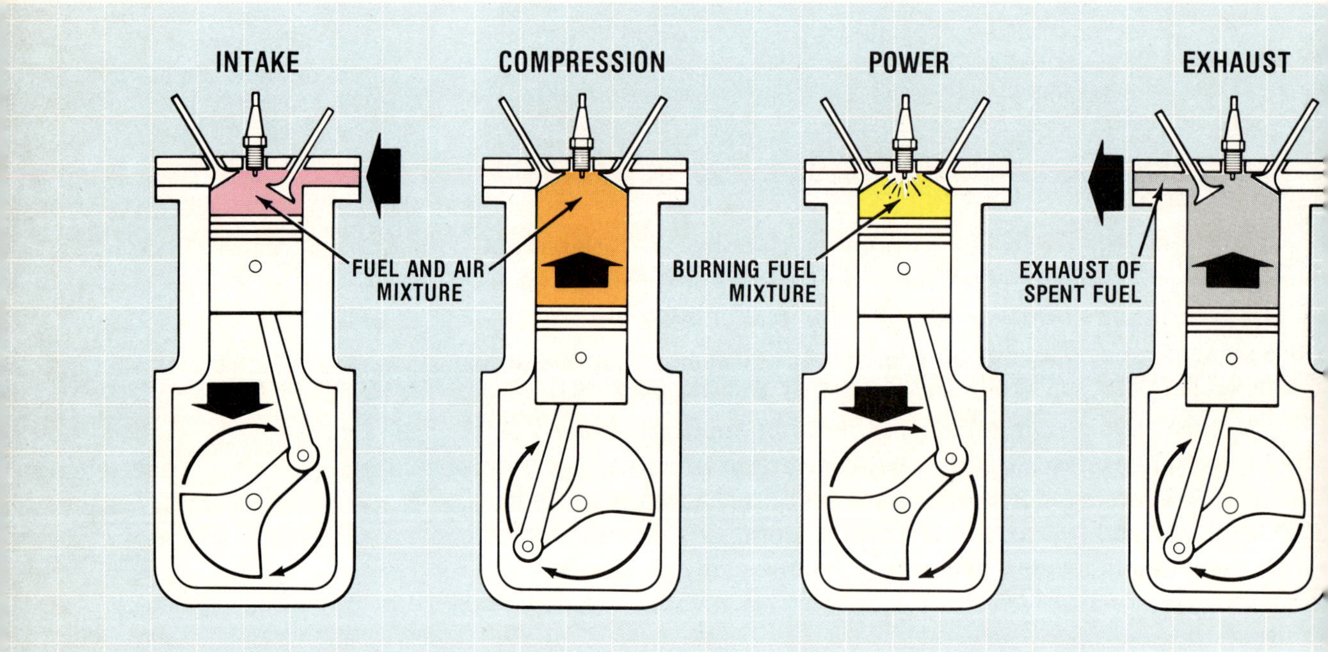


**The four-stroke cycle** is the most common means of deriving power from an internal combustion engine. The strokes, from left to right, are for intake, compression, power, and exhaust. These drawings show the functioning of a single cylinder. Automotive engines have multiple cylinders, usually four, six, or eight.

**Compression stroke.** During the *compression stroke,* the piston moves upward. This action squeezes, or *compresses,* the gas between the top of the piston and the head of the cylinder. Compression causes the fuel and air to mix thoroughly. Now the fuel and air mixture is ready for combustion.

**Power stroke.** When the fuel and air are fully compressed and ready to burn, an electric spark is created. This spark ignites (starts burning) the fuel by creating the needed temperature.

Combustion of the fuel is promoted by the supply of compressed air. Rapid burning causes explosive expansion of the gas within the cylinder. This expansion drives the piston downward in a *power stroke.* The power stroke actually delivers the work within an internal combustion engine.

**Exhaust stroke.** The *exhaust stroke* occurs with an upward movement of the piston. This pushes the burnt, or spent, gas out of the cylinder. The four-stroke cycle is now ready to begin again.

An internal combustion engine has multiple cylinders, usually four, six, or eight. The engine delivers power through a system under which four-stroke cycles are started at different times, in different cylinders. This means that one or two pistons are always in the power stroke. These power strokes keep the engine working while the remaining cylinders are at different points in their cycles.

## The Diesel Engine

The same four-stroke cycle as described above is used in *diesel* engines. As you may know, diesel engines are used widely. Diesels power automobiles, trucks, buses, trains, farm tractors, and irrigation equipment. Diesel engines are sometimes used to generate electric power.

Diesel engines use a fuel that has different burning properties than gasoline. Diesel fuel is made as part of the same petroleum refining process that creates gasoline. Diesel fuel is derived from crude oil before gasoline. Putting it another way, gasoline requires more refining than diesel fuel. One difference in the finished fuels is that gasoline is more *volatile.* Volatile is the property of a liquid for turning into a vapor, or gas. When a liquid vaporizes, the process is known as *evaporation.* Because gasoline is more volatile, it turns into a vapor at lower temperatures than diesel fuel. It is this property of gasoline that causes the fuel to explode when a spark is introduced into a cylinder.

Diesel fuel cannot be ignited in the same way as gasoline. A different method of starting combustion is used. Remember, the ingredients of combustion are fuel, air (oxygen), and temperature. In diesel engines, fuel and air are processed in a device

**Diesel engines** are used for many heavy vehicles, such as buses and trucks. Diesel engines operate without spark plugs. These engines are usually considered to be highly economical. (Queen City Metro, Cincinnati, Ohio)

known as an *injector.* An injector fills each cylinder with air. The air is then compressed by the upward stroke of the piston. As the piston nears the top of the cylinder, the injector feeds in a few drops of fuel.

Diesel engine cylinders do not use spark plugs for ignition. Instead, the temperature needed to support combustion is created by compression. A diesel engine achieves far more compression than a gasoline engine. As gas is compressed, its temperature increases. In diesel engines, the compression is great enough to cause the heated air and fuel to burn. Combustion results from compression. Therefore, no spark is needed for ignition.

## FUEL USAGE

All combustion-type engines, external or internal, share an important problem. That is, they require fuel for combustion. Fuel, in turn, is a critical natural resource. Critical means that fuel resources are important, vital. Critical also means that supplies of fuel are limited. There is only so much oil available in the Earth. Supplies have been great. But petroleum supplies are not endless. Sooner or later, the growing human population of the Earth will use up the available supplies of petroleum.

Another matter of concern is about the effects of combustion-type engines. The fuel that is used to drive these engines is converted to heat. In effect, the fuel goes up in smoke. Burnt fuel is no longer an energy source. Quite to the contrary, the burnt fuel is sent into the atmosphere—the space around the surface of the Earth. The smoke and particles of fuel become part of the air you breathe. Burnt fuel causes pollution. Pollution is one result of technology that is of concern to all people and to their governments. Concerns over fuel supplies and over pollution are important technology topics in themselves. These topics are covered elsewhere in this book.

## RELATED TECHNOLOGIES

To be useful, the technology of power needs other, related technologies. An engine delivers power. For this reason, engines are often called *power plants.* However, power alone doesn't provide a complete automobile. A number of technologies must work together within any device as complicated as a car, an airplane, or a computer. There must be *systems of technology.*

The technology of power within an automotive engine is described above. But creating work by driving pistons is only a small part of the technology needed in building an automobile. The automobile uses a number of separate technologies. These technologies form individual, related systems that function together within a car.

Consider why multiple systems are necessary. An internal combustion engine burns fuel. A system is needed to deliver fuel to the engine. Combustion in cylinders is created by electric sparks. It figures, then, that a car must have an electrical system. Considerable heat is generated. This heat could become so great that, if unchecked, the metal parts of the engine would expand and stick. The heat could even become great enough to melt some metal parts. Thus, a cooling system is required. A separate system is needed to direct the movement of the car. Other technologies are put to work to slow the car down and to give you a comfortable ride. Some of the related systems used in harnessing power within an automobile include:

- The *electrical system* delivers energy that produces the sparks needed for internal combustion. This system also provides energy for headlights, brakes, heaters, and other necessary units of a car.

- The *fuel system* includes a tank for storing gasoline. From the tank, the fuel system pumps the gasoline to the engine through tubes. A device called a carburetor, or an injector, mixes the fuel with air and delivers the mixture to the cylinders. Like the electrical system and others, the fuel system performs a separate, special operation. However, like the others, the fuel system is part of a single, overall system—the automobile.

**Multiple technology systems** are applied in complex equipment, such as the typical passenger car.

- The *cooling system* circulates air and fluid through the engine to keep it from overheating. This system includes a radiator that stores fluid and exposes the heated fluid to air for cooling.
- The *lubrication system* delivers oil to the moving parts of an engine. Lubrication means a spreading of oil to ease the movement of metal parts that rub together. Without lubrication, the moving parts of an engine would wear out. Lubricating oil reduces wear, helping engines to run smoothly and for a long time.

- The *steering system* makes it possible to turn the front wheels to control a car's direction of movement.
- The *brake system* slows the car down. When the driver steps on a brake pedal, pressure is applied to keep the wheels from turning.
- The *suspension system* gives the car a comfortable ride. A series of springs and shock absorbers cushion the body of the car. Passengers don't receive the full shock of bumps in the road.

## DIFFERENT TYPES OF POWER

Earlier, the internal combustion engine is described as an important source of power. This is true, particularly for jobs to be done on the ground. Internal combustion engines with pistons also power many boats and airplanes. However, power for flight also uses two other important types of engines:

- Turbine (jet)
- Rocket.

Both of these types of engines use *thrust* as the force to create movement. Thrust engines use the energy of expanding gas as a source of force. Air and fuel burned within an engine are forced directly out of the rear of the engine. This forced air, in turn, causes movement of an airplane or missile.

Turbine and rocket engines use internal combustion technologies. But power is delivered differently than from engines that use pistons.

### Turbine (Jet) Engines

The operation of a *turbine,* or *jet,* engine is diagrammed on the next page. A turbine is a device that rotates to deliver power. A jet is a forceful stream of air or water. The illustrated engine uses a turbine and delivers a jet of hot gases that thrusts the airplane forward.

Air is taken into a jet engine at the front. Incoming air is directed through a compressor, which squeezes, or concentrates, the air. The idea is similar to the compression of gases in a cylinder engine. That is, the purpose is to provide enough oxygen to promote rapid burning.

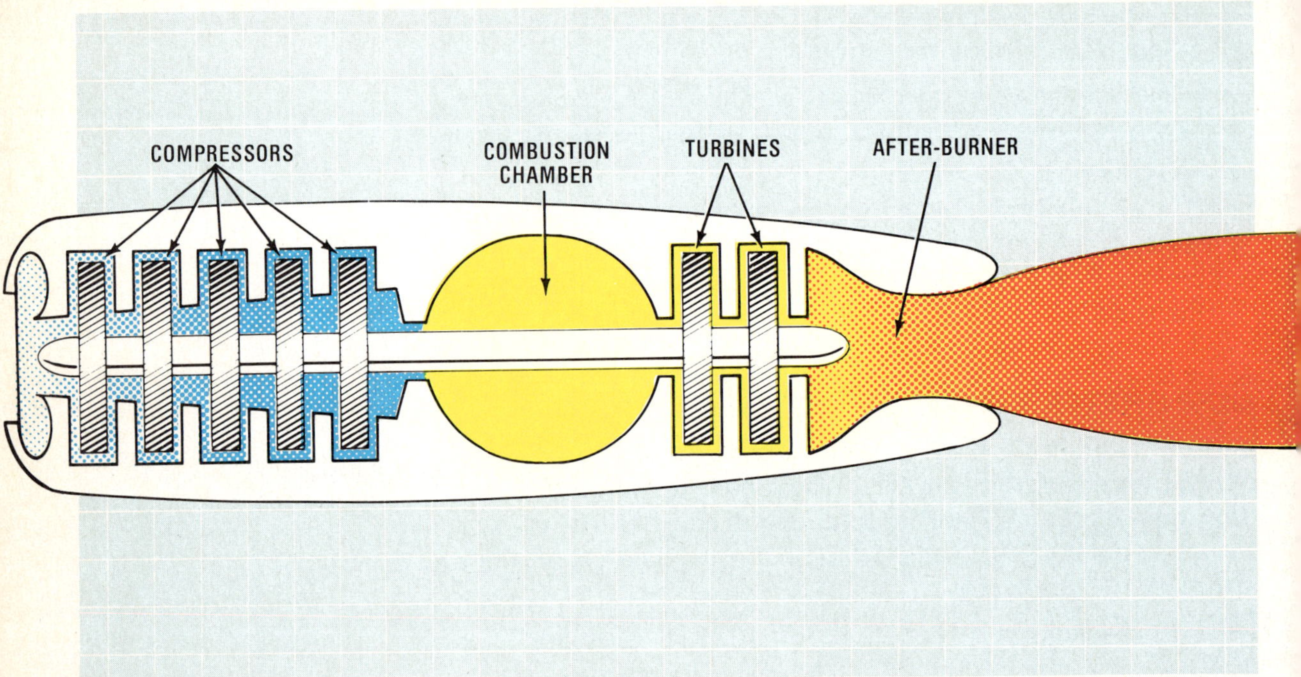


**Compression delivers power** for the flight of jet airplanes. This drawing shows the parts and functions of a typical aircraft jet engine.

As shown, the compressed air enters a combustion chamber. There, air and fuel burn rapidly, expanding through a turbine. The turbine drives the air out of the rear of the engine. This is the thrust that drives the plane. Note that the illustrated engine has a feature known as an afterburner. This applies additional heat to escaping gases, adding still more thrust.

## Rocket Engines

A *rocket engine* also uses thrust as its source of power. Rocket engines differ from turbine engines in the amount of power they deliver and in the way they use fuel. A diagram of a rocket engine is shown above.

One feature of a rocket engine is that it delivers more thrust than a jet engine of the same size. To illustrate, rocket engines power space vehicles and missiles. The thrust must be strong enough to drive the vehicle out of the Earth's gravity. An aircraft engine does not develop enough power to push a vehicle into space.

Another feature of a rocket engine is that it burns for only a matter of minutes. Special, high-temperature fuels are used. These burn so quickly that the supply is used up in minutes. By

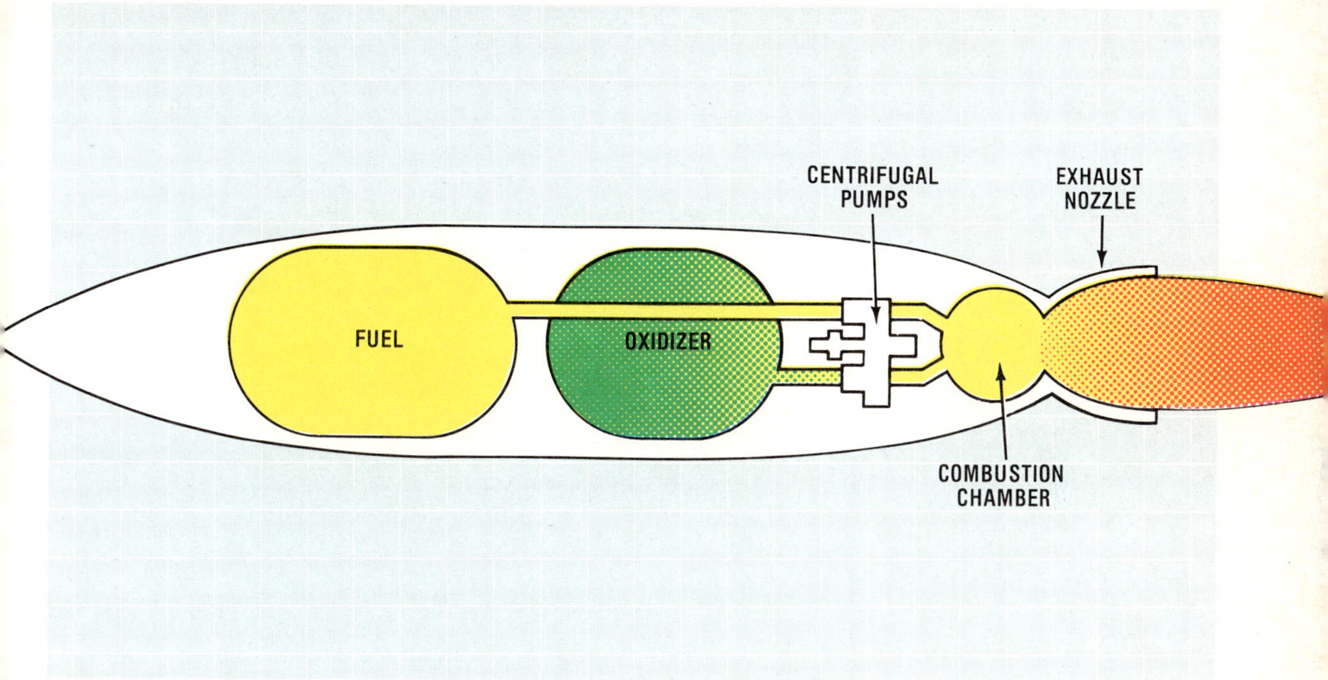


**Rocket engines** get their thrust from the expansion of rapidly burning fuel. Rocket engines deliver large amounts of power for a short time.

the time the fuel burns out, the space vehicle is in orbit. Then, other small thrust engines are used to guide an orbiting vehicle.

As shown, a rocket engine uses concentrations of special fuels and chemicals that supply oxygen (oxidizers). The oxidizers promote burning more rapidly than compressed air. The rapidly burning gases then provide the thrust for rocket-driven vehicles.

## RELATIONSHIP OF POWER TO ENERGY

Power, as discussed, uses sources of energy. The type of power described in this chapter can also be used to create certain types of energy. Internal and external combustion engines, for example, are used to produce electricity. Electricity, in turn, is the major source of energy for the modern world.

As sources of energy, power plants like those described in this chapter are often called *mechanical energy.* Mechanical refers to any physical device that moves. Thus, mechanical energy is the result of physical movement of powered devices, such as engines. The chapter that follows deals with the use of mechanical energy to produce a different type of energy.

## Summary

Certain terms have a precise meaning when related to technology. Power, for instance, is the rate of doing work. Work is the use of force to move something. Force is energy that has been put to use.

A power system combines many scientific and technological principles to produce sources of energy. Heating gas is commonly used in power systems. The power cycle of such a system has several parts: Fuel is burned to produce heat. Heat is applied to gas, which, in turn, expands. Expanding gas forces movement of some object (power results). Power is converted to a useful force.

One combustion system, external combustion, produces heat outside of the engine. The steam engine is an example of an external combustion engine. The external heat turns water to steam. The expanding steam becomes a source of power.

Another type of system is the internal combustion system. The internal combustion engine contains combustion within engine cylinders. The resulting heat expands gases to deliver power.

To have combustion, three requirements must be met: There must be fuel, oxygen, and a certain temperature.

Internal combustion engines use a four-stroke cycle in their operation. The intake stroke (of the piston) draws in gasoline vapors and air. The compression stroke squeezes and mixes the gas and air. The power stroke uses a spark to heat the mixture, which causes the gas to expand. Expanding gas moves the piston, and power results. The final stroke, or exhaust stroke, pushes the spent gas out of the cylinder.

The diesel engine also uses a four-stroke cycle. However, combustion is caused by the heat generated through compression of air. There are no spark plugs in diesel engines.

## Technology Terms

power
rate
work
force
power system
combustion
power cycle
external combustion
steam engine
boiler
cylinder
piston
piston rod
stroke
stationary

Relying on combustion engines presents two problems. One, fuel is required, and fuel supplies are limited. Two, burning fuel releases smoke and fuel particles into the air. The air, in effect, becomes polluted.

The technology of power must be combined with other technologies to be useful. For example, an automobile requires a number of systems of technology, including electrical, fuel, cooling, lubrication, steering, brake, and suspension.

Two internal combustion systems that produce force in ways other than through the movement of pistons are turbine and rocket engines. Both of these engines use thrust as the force to create movement. This means that the expanding gas is forced out of the rear of the engine, causing movement of the vehicle. The difference between a turbine and rocket engine is the amount of power delivered and fuel usage. A rocket engine uses a high-temperature fuel, which produces more thrust and burns quicker.

The power plants discussed in this chapter produce what is called mechanical energy, or energy produced by movement of a device.

## Using Technology

1. Suppose that you push against a concrete wall for four minutes. How much power have you delivered?

2. Suppose that your family drove a car to visit relatives. The round trip (to and from) is 240 miles. Gasoline costs $1.25 per gallon. Your family spent a total of $18.75 on gas for the trip. How many miles per gallon did the car get?

3. How are the cooling and lubrication systems in an automobile related?

flywheel
belt driven
internal combustion engine
fuel
oxygen
temperature
igniting
cycle
four-stroke cycle
intake stroke
compression stroke
compress
power stroke
exhaust stroke
diesel
volatile
evaporation
injector
power plant
systems of technology
electrical system
fuel system
cooling system
lubrication system
steering system
brake system
suspension system
thrust
turbine
jet
rocket engine
mechanical energy

# 6

# ENERGY

## Your Learning Job

After you have read this chapter and completed the work assignments, you should be able to:

- ☐ Define energy as it relates to technology.
- ☐ Describe energy states and tell how energy changes states.
- ☐ Explain the relationship between energy and power.
- ☐ Identify the source of electricity and explain how certain materials conduct electricity.
- ☐ Explain how electricity is generated by a battery.
- ☐ Describe how magnetism can be applied to generate electricity.

**Electricity is the main form of energy** produced by technology. The need for energy is illustrated well by this photo of New York City at night. (New York Convention and Visitors Bureau)

## THE NATURE OF ENERGY

Many people have experienced energy and its effects. Think back to a sunny day in your past. Perhaps you and friends decided to take advantage of the weather at the beach. Or maybe you spent the day at the pool or beside a beautiful lake. The sun seemed warm and friendly throughout the day.

Sunbathing was a pleasant experience. But, as night fell, things changed. Your skin began to feel chilled, then stiff. Pretty soon you were sore all over. It hurt to move. Your friends made bad jokes about your sunburn. You probably didn't realize it, but your sunburn was a personal, first-hand experience with natural energy.

Natural energy is a necessity. The energy from the sun is one of the important elements that supports life on earth. Without the sun, nothing could live. Plants, animals, and people all require the sun's energy. When you got sunburned, you simply had too much of a good thing. The heat of the sun supports life. But

**Sails in the wind** have been a symbol of energy for centuries. In this photo, the wind is driving a sailboat at Waikiki Beach, Hawaii. (Courtesy Hawaii Visitors Bureau)

too much of the sun's heat can damage and even kill the human body. To be useful, energy must be taken in limited amounts.

You have already learned about some natural sources of energy. The same energy that burns your skin can be used to warm water for solar heating systems. The sun can also generate electricity when *photovoltaic cells* are exposed to its light. A photovoltaic cell is an electronic device that converts light directly to electricity. In addition to the sun, natural energy sources include wind, running water, and the tides.

In technology, the term *energy* has a special meaning. This meaning is different from the way the word is used in everyday speech. For example, if you say a person is energetic, you mean that he or she is active. A person with energy might move around a lot and keep going without tiring. In technology, energy is any source of force. Another characteristic is that energy produces change. When the sun's *rays* shine on your skin, energy is changed to heat. The same rays cause a photovoltaic cell to produce electricity. Rays are the elements that make up the light that originates from the sun. These rays deliver the sun's energy.

Change occurs when running water turns a wheel that produces power to run machines. This change is from the energy of running water to the power from a moving wheel. Change also occurs when wind turns the vanes of a windmill. The power of combustion can become a source of energy. In this case, power is used to drive devices that produce electricity. The production, or generation, of electricity by power-driven devices is described later in this chapter.

These descriptions show that energy comes in many forms. These forms include wind, sun, water, tides, combustion, nuclear, and electrical energy. Energy, you will remember, is a source of power. To produce power out of energy, a change must take place. As change occurs, energy produces force, and force delivers work.

## STATES OF ENERGY

The energy that causes change represents one of three types, or *states,* of energy. Energy can change forms among these states. That is, energy in one state can take on another condition, or state. These three states are:

- Kinetic
- Potential
- Rest.

### Kinetic Energy

Change is caused by energy in action. This state, or type, is known as *kinetic energy.* Examples of energy in action, or in a kinetic state, include anything that has movement and can deliver force. A rock thrown by a person has kinetic energy as it travels through the air. A moving car or truck has kinetic energy. So does a piston inside the cylinder of an engine.

**Water in motion** can provide a source of kinetic energy. (Courtesy Hawaii Vistors Bureau)

Kinetic energy results from the movement of the source item. A car traveling down a highway develops kinetic energy from its own movement. If the engine were turned off, the weight and force of the car would still produce kinetic energy. Even if brakes are applied, the kinetic energy will keep the car moving for some distance. As long as water keeps moving, it applies kinetic energy to anything it touches. This is also true for a ball or a stone thrown through the air.

In its kinetic state, energy is actually being applied or used. Change is taking place.

**An example of potential energy** can be seen in water stored behind a dam. (Tennessee Valley Authority)

## Potential Energy

*Potential energy* is not actually being applied. Rather, the energy is ready for use.

As an example, consider the running water that turns a wheel. This is a form of kinetic energy. However, when water is held behind a dam, ready for use, it represents potential energy. A magnet that is always ready to attract iron materials is another form of potential energy. Your muscle, ready to go to work at your command, is still another form of potential energy. In all these states, energy is in a form that is ready for use.

## Rest Energy

*Rest energy* exists in *inert* objects that can deliver energy when force is applied. Inert means lifeless—not living and motionless. An example of rest energy is a rock lying on the ground. No energy is being delivered by the inert rock. Neither is there any potential energy because the rock itself can't cause its own movement. However, if you pick the rock up and throw it, energy is delivered. The weight of the moving rock generates a form of kinetic energy. Thus, an inert object that can be moved when force is applied is said to provide rest energy.

Other examples of energy at rest include coal or oil in the ground. These energy resources do not actually provide sources of energy in their natural state. Coal must be dug from the ground. Then heat must be applied before energy is released. Oil must be refined to produce sources of energy. Natural resources of this type can be thought of as energy in a state of rest.

**Energy changes states** when an object at rest is put into motion. A soccer ball lying on the ground is at rest. When kicked, the ball delivers kinetic energy. (© Jim Whitmer Photography/Nawrocki Stock Photo)

## Relationships Among Energy States

Energy in all three states can change according to conditions that exist or controls that are applied.

- In the ground, coal is energy at rest. Burning coal delivers kinetic energy.
- Water running in a stream delivers kinetic energy. When the water is stored behind a dam, it represents potential energy. Falling water can provide the kinetic energy to generate electricity. Sometimes, electricity is used to conserve the energy potential of water. For example, *hydroelectric* plants often produce electricity from running water. Often, there is electricity to spare at hydroelectric plants in the late night hours. When water is in short supply, this spare, or surplus, electricity is used to pump water back into the dam. This movement of water represents a change of kinetic energy back to potential energy.
- If a stone on a hillside begins to roll, its energy state is changed. The state of this energy source goes from rest energy to kinetic energy.

All energy, in all states, represents a resource that can be used to do work. People can control, or manage, changes in states of energy. This management should be aimed at getting the most benefits from available sources of energy. Managing changes in the state of energy sources is an important way in which people put energy to work.

## HARNESSING ENERGY

Energy and power are related closely. Energy provides force. When force leads to movement, work is produced. Thus, to do work, people need sources of energy that can be controlled. With controlled energy, it is possible to deliver power when and where there is work to be done.

In the early days of urbanization and industrial development, energy sources were limited. People depended on natural sources of energy or on converting the power of animals to energy. When wind didn't blow, ships didn't move and windmills didn't turn. In years when there was little rainfall, the water levels of rivers and streams were low. In really bad years, there might not be enough water to turn the wheels in factories. In short, natural sources of energy were not always dependable.

As described in the previous chapter, people gained some independence with the invention of the steam engine. Ships no longer depended on wind. Railroads were possible because

**Multiple types of energy** were used by some nineteenth-century ships. The vessel shown here uses both wind and steam as power sources.

of the power of the steam engine. Factories could be built at any location that had fuel available for engines. Soon, railroads became the main source of transporting coal. When this happened, factories could be built at any location served by railroads.

External and internal combustion engines were developed as sources of power. These engines powered ships, machines, and, later, trains. As people gained knowledge and experience with engines, they also learned that power could be used to develop energy. There is a two-way relationship: Power uses energy, but power can also produce new sources of energy. The increased energy, in turn, could be used to produce work, to deliver power.

In the modern world, power and energy have been brought together in the form of electricity. Electricity is the main source of energy used by people. The electricity on which people now depend is generated, in turn, by power systems. These power systems include widespread use of external and internal combustion engines.

## NATURE OF ELECTRICITY

*Electricity* is a form of energy. Understanding electricity and its sources requires, in turn, an understanding of the makeup of some natural materials. By comparison, other sources covered so far are easier to understand. The ability of running water to turn a wheel is easily understood. It is also relatively easy to understand how gases expand to drive external and internal combustion engines. These sources of energy exert force. With movement, the force is converted to work.

One problem with understanding the sources of electricity is that these sources can't be seen by people. So, special explanations are necessary. These explanations involve pictures of the materials that carry and deliver electrical force.

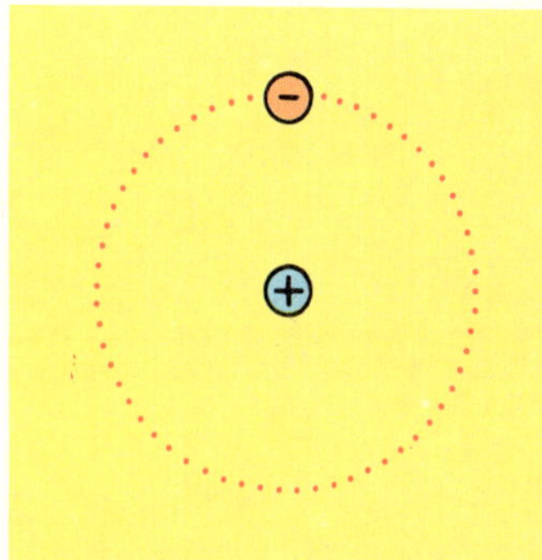

**Hydrogen is the lightest** of the atoms. As shown, a hydrogen atom has a single proton in its nucleus and a single electron in its outer shell.

Electricity originates and is carried in certain types of *matter.* Matter, in turn, includes all materials that make up the world and the universe. Your body is composed of matter. Water is matter, as are rocks and metals. All matter, in turn, is composed of *atoms* or combinations of atoms. An atom is the smallest, most basic building block for all materials. Atoms are so small that they can't be observed, even with microscopes. Rather, knowledge is built on deductions built from observations of the behavior of atoms and their parts.

To illustrate, look at the accompanying diagram of a hydrogen atom. Hydrogen is the lightest of the atoms that make up matter. In its natural form, hydrogen is a gas. Hydrogen atoms are part of all air and water. As shown in the diagram, the hydrogen atom is pictured through a series of circles and signs. At the center of the atom is a core, known as a *nucleus.* One of the parts that makes up the nucleus is a *proton.* In the diagram, the proton is represented by a plus sign (+). This indicates that the proton has a *positive electrical charge.* A charge is a unit of electrical force. A postive charge is one of two types of electrical force.

The second type of electrical charge is indicated by a minus sign ( – ). This identifies a *negative electrical charge*, which is the opposite of a positive charge. The diagram shows that the negative charge belongs to another portion of the atom, known as an *electron*. The dotted circle on which the electron is placed is called an *orbit*, or *shell*. The tiny electron circles the nucleus of its atom at extremely high speed. The electron is kept in its position in the orbit, or shell, by the force of its electrical charge. That is, the negative charge of the electron is attracted by the positive charge of the proton. This force keeps the electron rotating at approximately the same distance from the nucleus.

The attraction of the electron to the proton is extremely strong. The attraction is strong enough to keep the electron in its orbit despite the speed at which it travels. The speed of electricity is 186,000 miles per second. This is the same rate at which light travels through air. At this speed, even tiny, *subatomic* electrons exert great force. This force is the source of both electrical power and atomic (nuclear) energy.

Electrons are the source of electricity. Specific kinds of atoms are needed to conduct electrical power. A *conductor* is a material through which electricity passes easily. The makeup of these atoms is more complex than that of the hydrogen atom. To illustrate, look at the accompanying diagram of the copper atom. There are 29 electrons and 29 protons.

**The copper atom** has an ideal structure for an electrical conductor. The incomplete outer ring produces free electrons with relative ease.

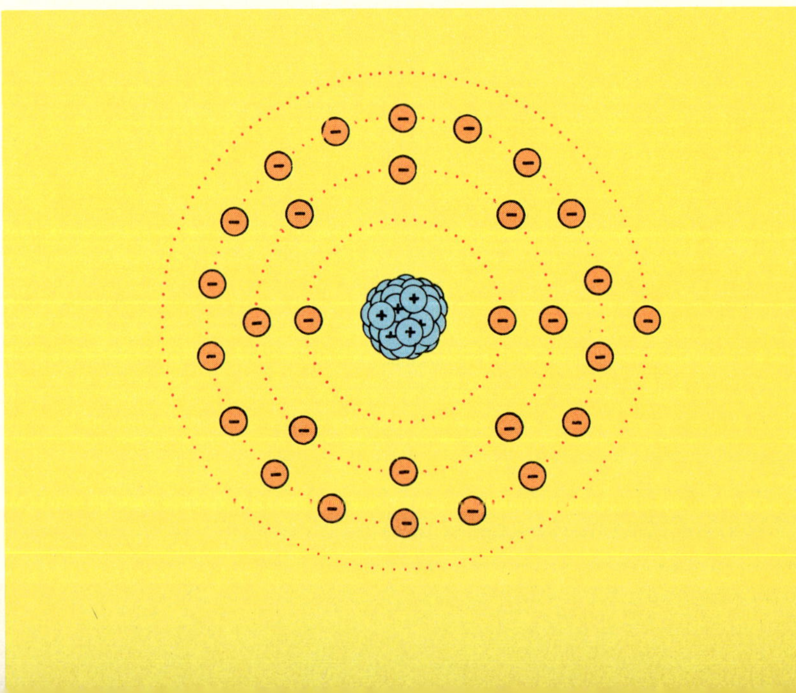

Note also that a copper atom has four shells of electrons. In all atomic structures, there can be no more than two electrons in the first shell. The second shell can have up to eight electrons. The third shell can have 8 or up to 18 electrons. The fourth shell can have 8, 18, or 32 electrons. The number of electrons that can fit into any given shell is the same for all atoms. The final shell in an atomic structure is called the *outer shell.*

When an atomic shell contains the maximum number of electrons it can hold, the atom is said to be *stable.* That is, the electrons in a full shell, also called a *ring,* are hard to loosen from their orbit. However, the shells in an incomplete ring are less stable. The *atomic bond* is said to be weaker for electrons in an incomplete shell. These electrons are often loosened from their outer rings. When this happens, the loose electrons are known as *free electrons.* Free electrons travel within the matter in which they are contained.

An atom that has fewer electrons than protons changes its electrical balance. Extra protons give atoms a positive attraction. Free electrons are attracted to these atoms. If an atom has fewer electrons than protons, free electrons can fill the outer ring. An accompanying drawing shows the movement of free electrons. Wire made of copper is a good conductor of electron flow. The electrons flowing through a conductor are known as an *electrical current.* An electrical current provides energy to produce work. Work produced from electrical energy includes light (in bulbs) and movement (motors).

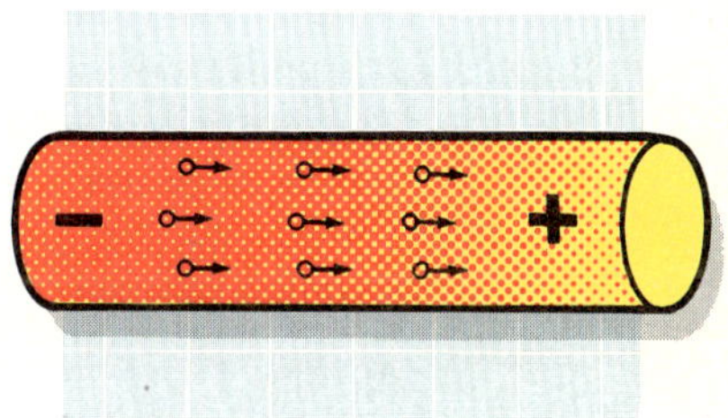


**Electricity flows** through conductors from the negative end toward the positive end.

Look at the drawing that shows electrons moving in a conducting wire. This indicates that the conductor has positive and negative ends, or directions. In this drawing, negative electrons are shown moving toward the positive end of the conductor.

Electricity, then, is a source of energy created by the flow of negative electrons. These electrons flow toward a positive attraction.

## ELECTRICAL BATTERIES

The attraction of electrons toward positive materials can be used to produce electrical sources, known as *cells.* Cells, in turn, are used to form *batteries.* A simplified diagram of a battery is shown on the following page. Each battery has positive and negative electrical sources, called *electrodes.* The electrodes are

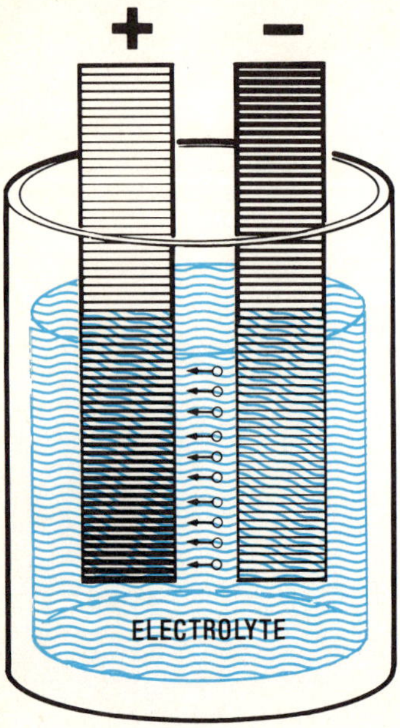


**A battery creates electricity** chemically. Electrons flow through an electrolyte from the negative pole to the positive.

placed in an *electrolyte,* which can be a liquid or a paste-like material. The electrolyte is a good carrier of electrons.

Within a battery, electrons move between electrodes. Conductors connected to the electrodes can receive electric current from the battery.

A battery that uses an electrolyte is a chemical generator of electricity. The electrolyte within the battery is the chemical. When the electron supply within a battery is used up, no more electricity can be generated. Therefore, a battery is limited in the amount of electricity that can be generated.

Batteries are good for jobs involving the generation of a small amount of electricity. However, modern society needs large amounts of electricity, which batteries cannot provide. Instead, technology has developed methods for generating electricity from mechanical power. The generators that use mechanical power are called *electromechanical generators.*

## GENERATING ELECTRICITY

Positive and negative forces that cause electrons to move exist in nature and in man-made devices. The devices, called *magnets,* create a force called *magnetism.* You have probably seen magnets at work since you were a small child. As you may know already, magnetism is a force that attracts, or draws, certain materials to a magnet. Materials attracted to a magnet contain iron, a metal element. Iron is an important building material.

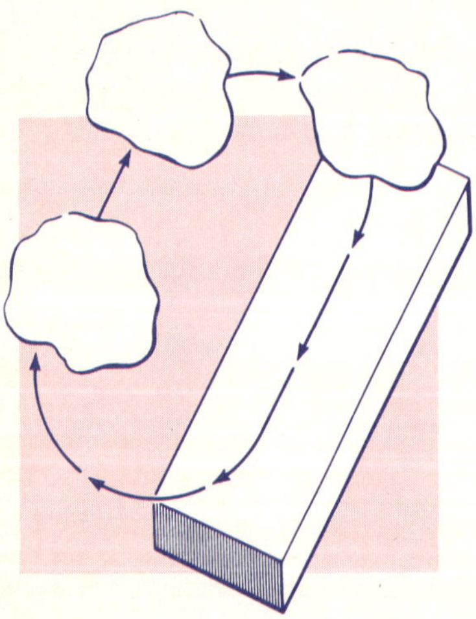

**A magnet can be created** by rubbing a lodestone over an iron bar.

Magnets that exist in nature are called *lodestones.* A lodestone is a natural rock-like material called ore. In its natural form, a lodestone will attract objects that contain iron.

Magnets can also be made by people. One way to make a magnet is to rub a lodestone against a bar of iron. Steel, an iron-based material, can also be used. This simple procedure is shown in an accompanying illustration. When a bar of steel is rubbed with a lodestone in this way, it becomes a magnet.

When a magnet is created, the atomic structure of the iron bar is affected. Atoms of iron are lined up facing in the same direction, as illustrated. An atom is the smallest part of an element that can exist alone. When iron is magnetized, the atoms are aligned in specific directions.

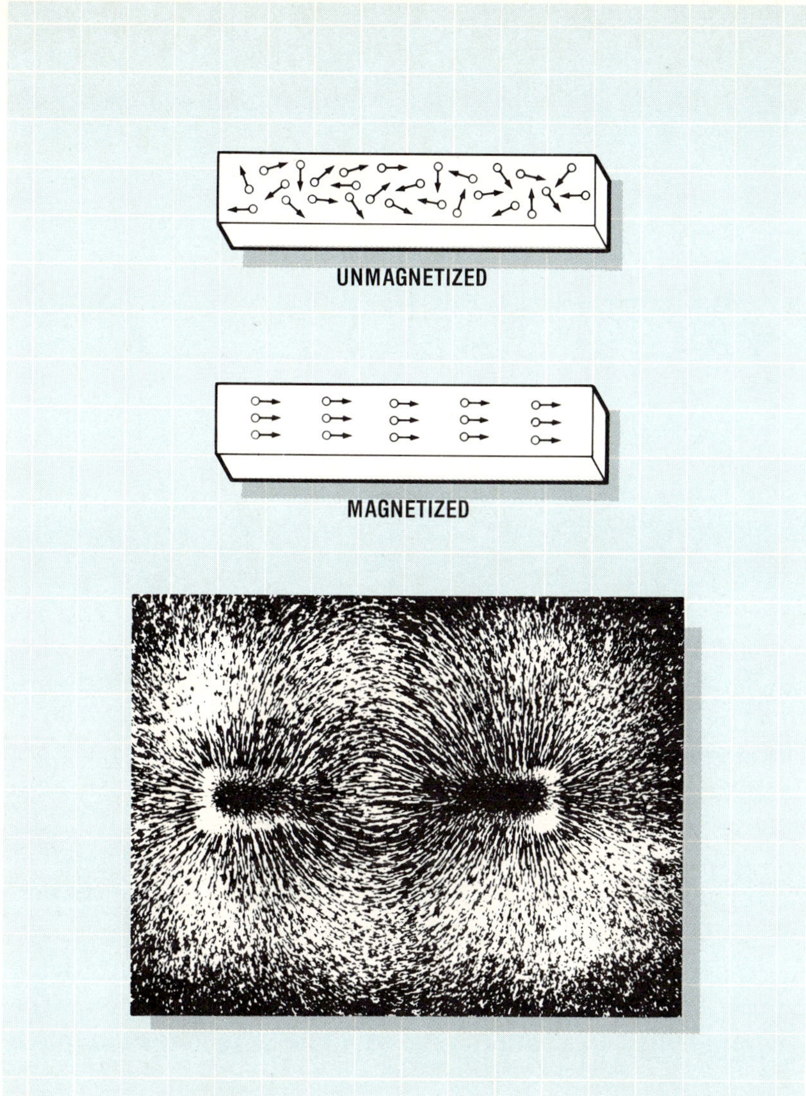


**Magnetic fields** occur when atoms of iron are magnetized and line up in the same direction. The lower illustration shows the effects of a magnetic field on a group of iron filings placed on a sheet of paper that, in turn, is placed over a magnet.

As shown, the atoms of iron line up along the length of the bar. All atoms point the same way. When this happens to the bar, the bar is called a magnet. This is because the atoms follow the force that causes magnetism. A bar magnet with its atoms facing in the same direction has a *magnetic field.* A magnetic field is a description of magnetic power. That is, magnetism reaches out to attract iron. This reach is the magnet's field. The field varies with the strength of an individual magnet.

A magnetic field can be demonstrated by placing a sheet of paper over a magnet. Iron filings are then placed over the paper. The filings line up according to the magnetic field of the magnet, as shown.

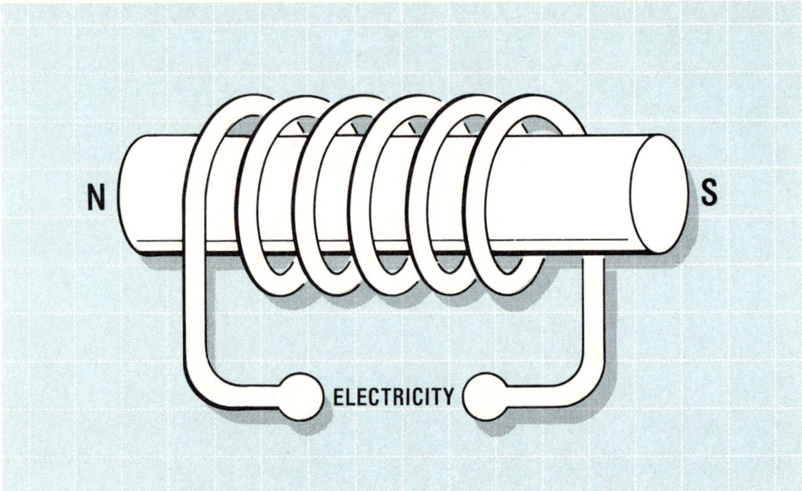


**An electromagnet is formed** by wrapping a coil of wire around an iron rod. When current flows through the wire, the iron is magnetized.

Magnetic fields always follow a set pattern. The ends of each magnet have their own magnetic properties. These ends are called *magnetic poles.* In this sense, the word pole means an end. If a magnet is allowed to swing free, it will line up with the magnetic poles of the Earth. Therefore, the ends of a magnet are identified as *north pole* and *south pole.* For this reason, magnets are used in instruments called *compasses.* These are devices that always point north-south. Compasses are used in setting courses for ships and for other travel guidance.

Magnetic force can be a source of electrical power. A technique for developing electricity from magnetism is shown in the accompanying illustration. A conducting wire is wound into a

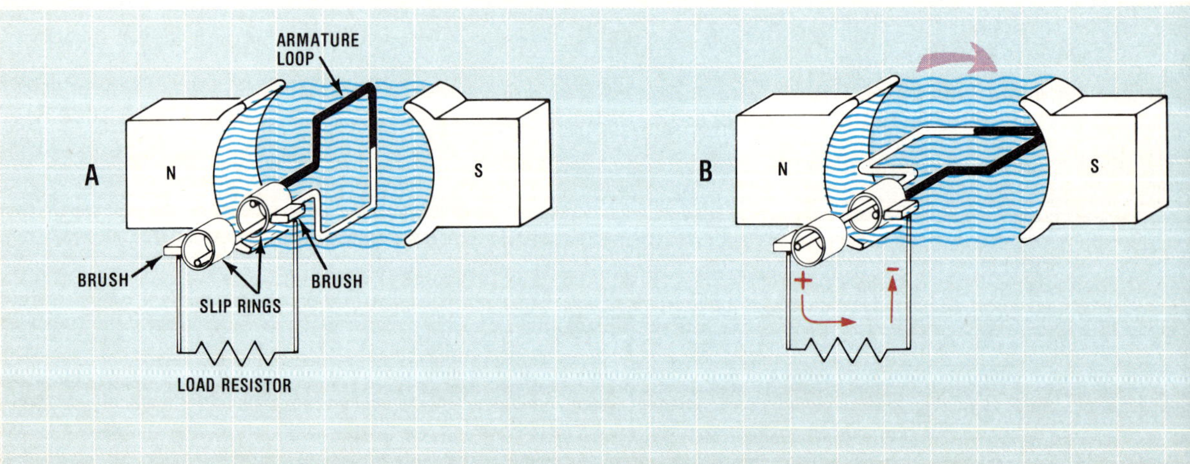

coil. When a magnet is moved through the center of the coil, the magnetic field causes movement of electrons. This movement of electrons becomes a flow of electrical current.

The relationship of magnetism to electron flow also operates in reverse. That is, a coil of wire can be wrapped around an unmagnetized bar. When an electric current is passed through the wire, a magnetic field is formed. Magnetism, then, results from the flow of electricity.

The relationship of magnetism and electricity is also the basis for developing electricity from power. The generators that produce electricity for your home and school use this technology. The accompanying illustrations show, in a simplified form, how power is generated. The parts of a generator are identified in drawing A. These are:

- A magnetic field is created, as shown by two separate poles with magnetic lines between them.
- An *armature loop* is placed within the magnetic field. An armature is a metal frame around which coils of wire are wound. The armature, then, provides a coil through which electrons will flow. To cause the flow of electrons, the coil must be moved within the magnetic field. This movement occurs with the rotation of the armature loop.
- *Slip rings* are holding devices for the armature loop. These rings position the armature loop and permit it to turn.
- *Brushes* are electrical devices. They carry current from the armature loop into the wires that deliver electricity.

**Electricity is generated** when an armature loop is passed between poles of a magnetic field. The armature loop turns continuously. When the armature loop is in an upright position, as in drawings A and C, no current flows. When the armature loop is horizontal, as in drawings B and D, electrical current is generated.

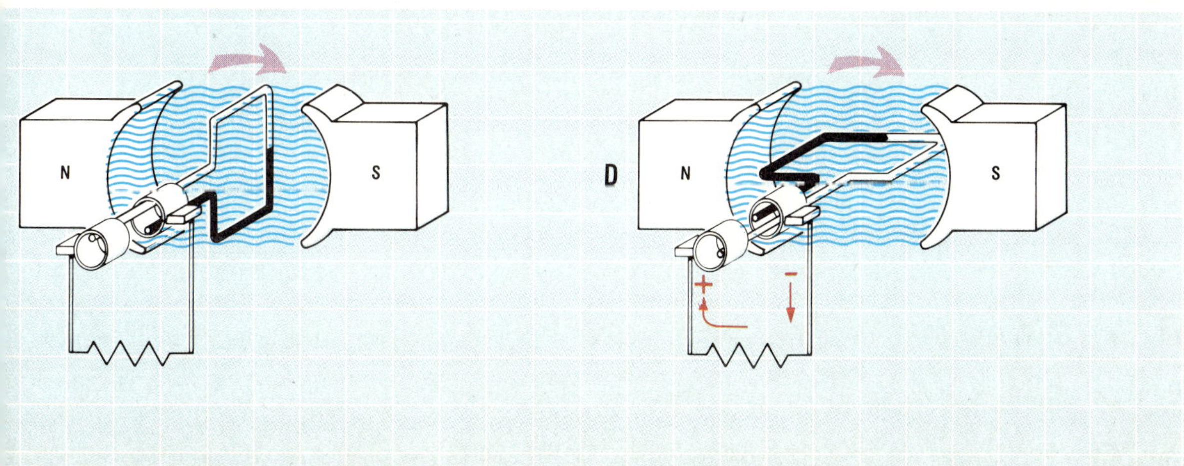

**Electric energy** can be converted to light through use of bulbs that have resistance wires. The resistance causes the wires to heat and give off a bright, white light. Lighting is one of the main uses of electrical generation.

- The *load resistor* represents the use of electricity. *Resistance* is the slowing down of the flow of electrons. Materials used in electrical equipment handle current differently. A conductor lets current flow freely. A resistant material holds back, or obstructs, the flow of current. When electric current is resisted, light and heat can be produced. The wire in a toaster, for example, is a resistant material. Special resistant wires are also used in light bulbs. In the generator diagram, the load resistor represents all uses for the electricity that is produced.

Note that electrical generation patterns change as the armature rotates. When the armature loop is pointed straight up and down, no electricity flows. Current does flow when the armature loop is directly across the magnetic field. The arrows pointing to and across the load resistor indicate directions of current flow. The direction of current flow is reversed as the armature loop turns.

The electricity produced from this generation technique is known as *alternating current (AC)*. To alternate means to change regularly. Alternating current changes its direction of flow on a regular basis. AC is the most commonly used form of electric power.

In reality, electrical generators are large devices occupying complete buildings. Power from water or from external or internal combustion engines or nuclear heat generators is used to drive the huge generators. Electricity is then carried through wires that deliver its energy to points of use. Electricity can travel hundreds of miles between generators and users.

## THE POWER-ENERGY CONNECTION

Power comes from energy. But energy is also a source of power. Both power and energy are needed to support a modern society.

Electricity, in particular, has become a necessity of modern life. This is demonstrated when there is a failure in an electric power system. Any major power failure is regarded as a disaster. Elevators don't operate in tall buildings. If the failure is at night, entire cities are left in darkness. People can't heat their homes. Many can't cook their food.

Looking at it another way, electricity is one of the things you count on in everyday life. When you operate an electric switch,

**Distribution of electricity** is demonstrated in these photos. The top photo shows part of a generating station at which electricity is created. Electricity is then carried over power lines, above, and delivered to users, as represented by the homes in the photo at right. (All photos, courtesy American Petroleum Institute)

you expect lights to come on. If you press a button in an elevator, you expect to be carried to the floor you select.

Perhaps more than anything else in modern life, electricity demonstrates the importance of technology. Electricity is part of virtually all modern systems of technology.

## Technology Terms

photovoltaic cell
energy
ray
state
kinetic energy
potential energy
rest energy
inert
hydroelectric
electricity
matter
atom
nucleus
proton
positive electrical charge
negative electrical charge

## Summary

In technology, energy is any source of force. Natural sources of energy include the sun, wind, and running water. Through technology, people have learned to control and generate energy.

Energy exists in one of three states: kinetic, potential, and rest. Kinetic energy is energy in action. An example would be a moving truck. Potential energy is ready for use but not being applied. A magnet, ready to attract iron materials, is an example of potential energy. Rest energy is found within inert objects that can deliver energy when force is applied. A rock lying on the ground illustrates energy at rest.

Energy can change states, depending on the conditions. For instance, coal under the ground is energy at rest. But when the coal is mined and burned, it represents kinetic energy.

Energy and power are closely related in that energy provides force. When force causes movement, work is done. To do work, then, people need to control energy.

Electricity is the major source of energy in a technological society. To understand how electricity works, you must understand the nature of matter. Matter is composed of atoms. At the center of every atom, there is a nucleus. Protons within a nucleus are attracted to the electrons, which orbit the nucleus at high speeds. If the attraction between the protons and electrons is balanced, the atom is stable. If, however, there is an imbalance, free electrons will travel through matter to correct the imbalance. The movement of negative electrical charges (electrons) provides the source of electrical energy. The electrons moving through a conductor are known as an electrical current.

Batteries are chemical generators of electricity. Electrodes (positive and negative electrical sources) are placed in an electrolyte, a good conductor. Batteries are useful for jobs that need only small amounts of electricity.

For larger amounts of electricity, generators are used. Generators use magnetic force to cause electrons to flow. A magnet is an object that attracts iron. Magnets are surrounded by force fields that set up attraction for iron. When a coil of conducting wire moves through a magnetic field, electrons flow in the wire. When an armature loop is rotated through the field, electrons flow through coils. Brushes carry the electric current from the armature loop to wires. Because the armature loop's position is changed, or alternated, the type of electricity produced by this generation technique is called alternating current (AC).

## Using Technology

1. List several examples of kinetic, potential, and rest energy.

2. Make a list of the appliances and devices in your home that require electricity. Which ones do you think are necessary? Which ones are luxuries?

3. A utility company measures energy in kilowatt-hours. A kilowatt-hour is equal to 1,000 watt-hours. Suppose that the charge for a kilowatt-hour of electricity is $.03. How much would it cost to burn a 100-watt light bulb for 10 hours?

electron
orbit
shell
subatomic
conductor
outer shell
stable
ring
atomic bond
free electron
electrical current
cell
battery
electrode
electrolyte
electromechanical generator
magnet
magnetism
lodestone
magnetic field
magnetic poles
north pole
south pole
compass
armature loop
slip ring
brush
load resistor
resistance
alternating current (AC)
network

# 7

# STRUCTURES

### Your Learning Job

After you have read this chapter and completed the work assignments, you should be able to:

- ☐ Trace the development of structures, from early caves to skyscrapers.
- ☐ Name the basic shapes that form most structures.
- ☐ Describe the strengths and weaknesses of each shape when used in structures.
- ☐ Describe the problems people overcame to develop skyscrapers.
- ☐ Describe perimeter-supported structures and explain their uses.
- ☐ Describe two methods for designing bridges.
- ☐ Explain how shell structures are suitable for airplanes and automobiles.
- ☐ Explain what is meant by the rule "Form follows function."

**A structure of beauty and fame** is the Parthenon, built on top of the Acropolis, a hill in Athens. (Greek National Tourist Organization)

## MEETING THE NEED FOR SHELTER

Shelter is one of three necessities of human life. The others are food (including water) and clothing. The basic need for shelter is protection. Shelter protects people from the damage that can be caused by weather or by natural enemies (animals).

### Caves

To meet the basic need for shelter, many early people lived in caves. Once people learned to use fire, the caves became more comfortable. However, caves were never perfect as shelters. One problem was that caves were formed by nature. People had no control over the size or shape of cave shelters.

Problems also developed with food supply. Early people were hunters and gatherers. Parties of hunters would leave caves

**The first shelters** sought by ancient people were caves. (Courtesy Jordan Information Bureau)

in search of animals. Dead animals would be brought back for food. When a group of people lived in a cave for a while, animals sensed the danger. The animals moved on to other areas—seeking places where they wouldn't be hunted. People had to keep moving in search of new animals to be killed for food.

As people learned to use plants for food, the same problem would dcvclop. Groups of pcoplc would cat up tho food supply in an area. It became necessary to move on in search of new food sources.

## Movable Homes

As they moved about, people developed portable shelters that they could take with them. Poles were cut from trees, leaned toward each other, and tied at the top. Then animal skins were stretched over the framework of poles. These shelters could be taken apart quickly and carried from place to place. People could travel anywhere in search of food, taking their shelters with them.

**Movable homes** became necessary as people traveled in search of food. Typical movable structures are represented by American Indian tepees. (© 1981 Bill Staley/West Stock)

## Permanent Homes

Eventually, people learned to farm. Once they could grow their own food, they didn't have to move around to keep eating. People could build permanent homes. The first permanent shelters were made from sticks and mud. Huts were built to shelter both people and farm animals. Gradually, different materials were used. Rough huts were replaced by structures such as log cabins. Eventually, people developed tools that made it possible to cut and finish wood. Wood-frame and stone houses were constructed.

## Urbanization

As people built and occupied towns and cities, their needs for structures changed and grew. In cities, many families might live in a single structure. It became necessary to build living structures with multiple floors, or *stories*. New building technologies were developed. Structures had to support multiple layers of rooms. Staircases were designed to enable people to move from one floor to another. Larger buildings required different bases, or *foundations,* than single-story dwellings.

Entirely new types of structures were needed for the factories around which cities developed. Factories needed open spaces as work areas for hundreds, even thousands, of persons. Structural technologies had to be developed that could provide these open areas.

The designs for factories had some features in common with the large churches and cathedrals of major cities. In a cathedral, structures had to be designed to permit many people to attend. All people had to be able to see and hear what was happening. Again, the need was for large, open areas.

Thus, structures became more than basic necessities that kept people warm and free from danger. Structures had to become places where people could work, worship, and gather for other purposes. Structures are results of technologies that have, in turn, supported the development of societies. Structures make modern life possible.

## STRUCTURAL FORMS

The technologies that have developed to meet human needs use certain basic shapes. These shapes are used individually and together. The structures that use these shapes or combinations of shapes are built around a few simple forms. These are:

- Triangle
- Circle/arch
- Square/rectangle.

### Triangle

A *triangle,* or *triangular shape,* is a form with three sides. The first use of the structural support available from a triangle may well have been in early tents. Tents, remember, had a series of poles as a *framework.* A framework is the basic structure that holds up a building. Frameworks are covered with other materials that separate the inside of the structure from the outside environment. Early tents, for example, used animal skins to cover the basic structure, or framework.

Building a tent similar to tepees used by American Indians begins with a single triangle. Other poles are then added to form a *free-standing* structure. A free-standing structure is one that supports itself. As indicated, a free-standing structure can be created by setting up multiple triangles.

In the example of the tent, note that two sides of the triangle form the actual structure. The earth provides the third side, or *base.* A base is necessary for any building that will stand in a fixed position.

The triangle has been a common form in structure technology through the years. For example, in ancient Egypt, structures were built that combined four triangular shapes. The result was a structure called a *pyramid.* The pyramids of Egypt illustrate how the basic triangular shape can be adapted to produce

**Triangular shapes** have been adapted for many buildings, including religious centers like the Mormon Tabernacle in Salt Lake City, Utah, left, or an oriental temple like the one shown above. (Photo at left, courtesy U.S. Department of Housing and Urban Development; at right, UN Photo)

a strong structure. The pyramids in Egypt have stood for thousands of years.

The triangular shape has also been popular as part of many religious buildings. Many cathedrals and churches have triangular-based roof structures called *spires.* A spire is a pointed roof structure. Spires are usually tall and narrow. In a structural sense, spires are usually decorative. That is, spire structures don't cover much area—considering the amount of materials used. However, spires help identify the structures on which they are built. Customs that have developed through the years identify spires with churches or cathedrals. In addition to providing shelter, the structure of modern buildings can also indicate their purpose.

The triangular shape is used commonly in the roof structures of many of today's homes. One typical form used on many

**The slanted roof, or gable,** is a basic use of the triangular form in structures. An advantage of the gable is that rain and snow slide off the roof.

**Triangular shapes are also featured** in the three structures above. The log cabin has a slanted roof to shed snow. The covered doorway in the center photo protects people entering and leaving a home. A dramatic example of a triangular shape can be seen in the San Francisco skyscraper at right, the Transamerica Pyramid. (Photo at left, Montana Travel Promotion; at right, courtesy U.S. Department of Housing and Urban Development)

homes is the slanted roof. The roof of a building comes to a point, or ridge, at the top. The roof then slopes off to either side of this high point. When the sides of a roof slope away from a high point, the design is called a *gable*. Gabled roofs are used in many areas that have cold winters. Snow slides off a gable structure. Without gables, the roof could cave in from the weight of the snow. The gables in a roof form triangles with *rafters*. Rafters are flat structural boards that complete the triangular shapes of gabled roofs.

Even tall buildings sometimes use triangles as their basic design. Recently, a number of tall buildings in large cities have been designed in the shape of tall, thin pyramids.

## Circles and Arches

Triangles provide a convenient way to design *portable* structures. Portable means movable. Tents are portable because they can be taken down, carried to new locations, and put up again. As people tamed animals and became farmers, however, it was no longer necessary to keep moving in search of food. People wanted *permanent* homes. Permanent homes last for many years and are not portable.

When it comes to building permanent homes, triangle structures can cause problems. Structures must be relatively high in relation to the living area made available. Think about a tepee. Long poles are used. The resulting structure is high and narrow, leaving not much room to move around inside. Also, a triangular shaped structure (or pyramid) has sloping sides. The areas near the edges of the insides of these structures are too low for people to stand. Thus, triangular buildings are high structures that provide relatively small interiors.

As people thought about permanent buildings, they undoubtedly wanted more space. People also wanted to develop materials that were easier to handle. In some areas, there were no tall trees available to use as poles. In these regions, people had to look for other ways to support structures. The answer came through use of *flexible* materials for structures. Flexible means that materials bend. The materials used included small bushes or reeds. A reed is a hollow, grass-type plant that bends easily. With these flexible materials, people began to build structures in the form of *circles* and *arches*. A circle is a perfectly round form with a uniform *radius*. A radius is the distance from the center of the circle to the outer edge. In a circle, the radius is the same for all points around a circle. An arch is a portion of a rounded shape. One form of arch is a half-circle. Arches can be used as the basic form for structures or as supports for structures.

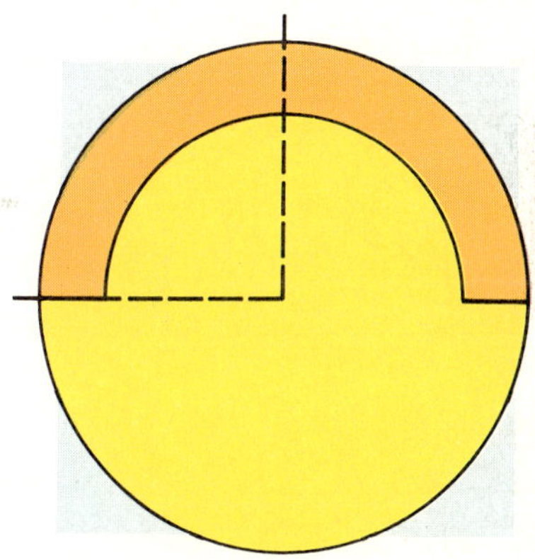

**The circle and partial circle, or arch,** provide basic forms used in many structures.

People used branches from small bushes and reed-type plants for early forms of permanent structures. The branches and reeds were woven into a *framework* for small homes. A framework supports a structure and gives the structure its basic shape. On top of the framework of early permanent homes, people spread mud. The type of soil used was called *adobe*. Adobe soil contains large amounts of clay. Buildings made from this material are called adobe structures. When adobe mud dries,

**The circular building form** was used in the hogan, a common type of structure used by Indians in the American Southwest.

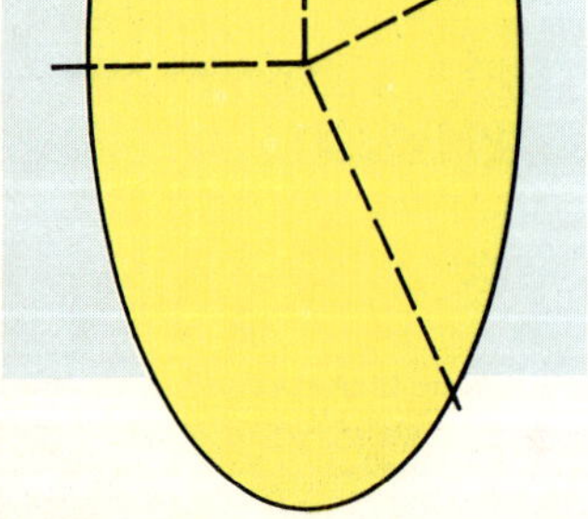

**A variation upon the circle** as a building form is the ellipse. The ellipse also can be used in part to form arches, as shown in the bridge on the opposite page.

the framework is strengthened to form a solid structure. Adobe structures were built by some American Indian tribes in the Southwest. The homes built in this way were called *hogans.*

Hogans were round structures with rounded roofs. These structures were relatively easy to build. Hogans also provided relatively large amounts of living area.

Arches have also played an important role in the development of structures. Many buildings in ancient Rome used arches in their design. This is partly because the arch has a high amount of structural strength. For this reason many bridges, right up to modern times, use arches.

Circular structures have special advantages in many modern buildings. As one example, a number of hospitals built recently use circular shapes. In round hospitals, nurses and other health-care professionals work at central locations. This provides quick access to the rooms of patients who require attention in a hurry. Also, in a circular structure, nurses can accomplish more patient care with less walking.

Not all round structures are circles. Another rounded form often used for building structures or arches is the *ellipse.* An ellipse is a rounded form that does not have a uniform radius. That is, an ellipse may have varying measurements at different radius points.

**Circular forms** are used in a wide variety of structures, ranging from grass huts, to stone arches, to modern sports stadiums, as well as for bridges. (Photos, from top: courtesy of the South African Tourism Board; © 1981 Spenser Church/West Stock; Queen City Metro, Cincinnati, Ohio)

## Squares and Rectangles

Triangular and rounded structures were easy for people to build in the days before tools were invented. It was simple to lean poles or trees together and to tie them in place. It was also simple to weave a structure from branches and reeds, then apply adobe mud. As people developed tools and building materials, such as nails, it became possible to add dimensions to structures.

In time, structures were built with four or more sides. These buildings took on different shapes and features. Doors were added to make it easier to get in and out. Different structural shapes were combined as sloped roofs were put on buildings with straight sides. Fireplaces were designed and became popular. With fireplaces, heating and cooking became more convenient.

These new structure types use two important shapes—the *square* and the *rectangle.* A square is a form that has four straight sides of equal length joined at *right angles.* A right angle has

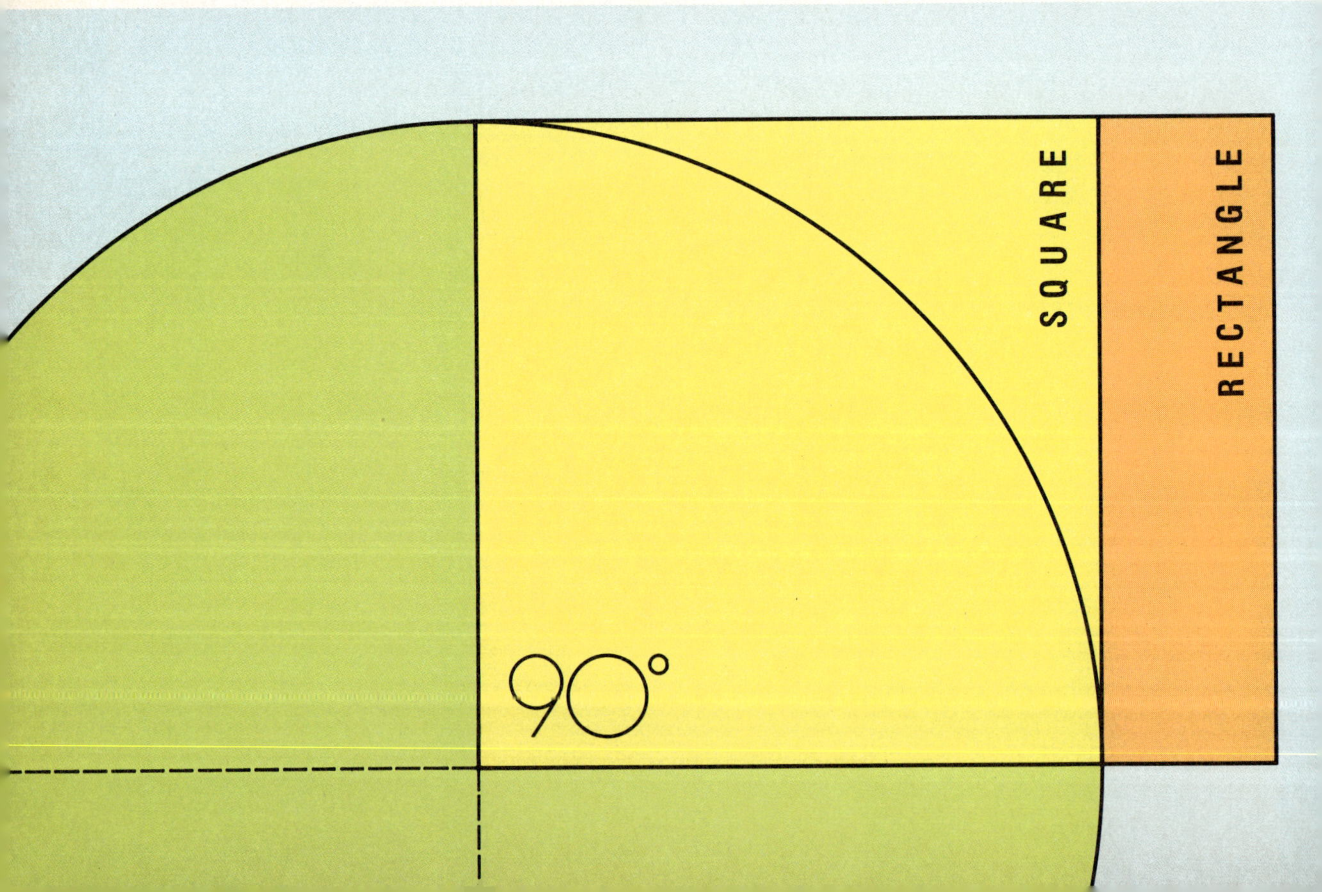


**Squares and rectangles,** basic shapes in modern structures, are shown in relation to a circle in this diagram. Note that both the square and rectangle have four straight sides joined by right angles.

sides joined at 90 degrees. A right angle occupies one-quarter of a full circle (360 degrees). A straight line marks off half a circle, or 180 degrees. So, a right-angle line marks off half the distance between the ends of a straight line. Lines at right angles are said to be *perpendicular* to each other. Perpendicular means that a line cuts straight across another line. Right angles and squares are illustrated in the accompanying diagrams.

A rectangle also has 90-degree angles and four straight sides. However, in a rectangle, all four sides do not have to be equal. Instead, there are two sets of *parallel* lines, with each set equal in length. Parallel lines are drawn at equal distances from one another over their entire length.

Square and rectangular structures became possible as people developed special building tools and techniques. Advanced building methods made it possible to join long wood or metal parts, called *beams,* at right angles. The beams then formed the *framing,* or framework, for structures. Today, most rooms, in most buildings, are rectangular or square.

**Square and rectangular shapes** form the basis for the exterior design of this modern office building.

**The pentagon and the trapezoid** are basic shapes used in many buildings. The photo above shows the Pentagon, headquarters for the U.S. Department of Defense. At right is a modern building with a trapezoidal shape.
(Photo above provided by Uniphoto)

In addition to squares and rectangles, other straight-sided shapes have also come into use. Some other shapes that are used regularly include:

- A *trapezoid* is a four-sided shape formed from lines that may be unparallel. At least two sides of a trapezoid are unparallel. Angles of a trapezoid may be unequal. One commonly used trapezoidal shape is formed, in effect, by removing the pointed top of a triangle. Trapezoids can provide great strength in designing structures.
- A *pentagon* is formed from five straight lines. A famous building of this design is The Pentagon, headquarters for the U. S. Department of Defense.
- Other straight-sided forms include the hexagon (six sides) and the octagon (eight sides).

## FRAMING

Framing supports a building. The term *framing* also describes special elements, or parts, of a structure. In a modern building, the framing consists of vertical (straight up and down) and horizontal (flat) beams. These beams are joined to form a framework that holds up the building. The framework of a building is similar to the skeleton of a human body. The skeleton holds up the muscles and organs of your body. Likewise, the framing of a structure holds the covering for the walls and the outside (exterior) surfaces. In a building with multiple stories, the lower levels of framework support the upper levels. For all structures, the framing for the walls supports the roof.

**Framing provides the basic support** for and shape of many buildings. This photo shows the construction of a wood frame for a typical private home. (American Petroleum Institute)

## EXTERIORS

Exterior (outside) walls or wall coverings of buildings can serve a number of purposes, including:

- Exterior coverings—including outside walls, windows, and roofs—protect against the weather. The people and furniture in a building are kept warm and dry because of protection from exterior coverings.
- Exterior coverings give the building its overall appearance.
- Exterior walls can provide structural support for the entire building.

Many different materials are used for exterior walls and coverings. These include steel, aluminum, cement (concrete), wood, glass, and shingles. Shingles are units of covering material that overlap as they are attached to a building.

**Skyscrapers conserve land,** particularly in large cities. This photo shows the Sears Tower, the world's tallest building. (Courtesy Sears, Roebuck and Co.)

## HIGH-RISE BUILDINGS

Structures use land. The more people there are in an area, the harder it becomes to find land for new structures. In large cities, land becomes both scarce and expensive. The cost of land goes up because of high demands for more buildings.

As cities developed, it became apparent that buildings would have to be built higher. This was the only practical way to provide housing and work space for the millions of people in large cities. Until the 1890s, there was a limit to how high buildings could go. The problem was that people could only climb so many stairs. After that, they were too tired for work or home activities. In most cities, practical limits were established for buildings with staircases. This limit was approximately five stories (floors).

But more living and work space was needed. Structural designers figured out how to build frameworks that would support higher buildings. Early *high-rise* buildings, or *skyscrapers,* depended heavily on exterior walls for support. Some early skyscrapers had exterior walls that were three or four feet thick at the ground floor. At higher levels, wall thickness was reduced.

Structural design was only part of the problem, however. People learned to construct tall buildings, but there was still a problem of how to transport people up and down. The

**The New York City skyline** is famous for its high-rise buildings, including the World Trade Center (left) and the Empire State Building. (Both photos, courtesy New York City Convention and Visitors Bureau)

skyscraper illustrates how systems of technology develop. Structural design alone couldn't provide the needed living and working space. The breakthrough came with the invention of the elevator. The idea was to transport people up and down with powered equipment. Trains had already come into use to carry people from one town to another. Elevators were developed as means of transporting people from one floor to another. Almost overnight, high-rise construction dominated American cities. Today, there are many buildings that have 100 or more stories.

**Perimeter-supported structures** are used for buildings in which large clear spaces are needed inside, as is the case with Durham Cathedral, England. (British Tourist Authority)

Today, it is no longer necessary for skyscrapers to have extra-thick walls. Instead, the interior framework of a modern high-rise building bears most of the weight. Frameworks consist of steel beams that hold the weight of the building from the inside outward. The even spreading of weight has made today's extra-tall buildings possible. Also, when strong interior framing is used, it becomes possible to use light exterior covers. Many modern skyscrapers have exterior coverings that consist largely of glass or aluminum panels. Such exteriors are possible largely because the weight of the building is supported by a strong interior structure.

## PERIMETER-SUPPORTED BUILDINGS

Not all of the needs for structures can be met by tall buildings. Another structural need occurs when many people have to gather or work in the same place.

To illustrate, consider structures built entirely for entertainment—theaters, sports arenas, and auditoriums. Many people gather in these buildings to watch performances or to take part in events. The seating arrangement must allow each person a clear view of the event. Also, extra room is needed so that people can move around. This means that the buildings need large amounts of clear area. There can be no beams or posts to support the upper stories or ceiling. Instead the ceilings are supported by large rafters or special frames. These structures are said to be *perimeter supported.* Perimeter refers to the outer edges of a place or structure.

Perimeter-supported buildings have strong exterior walls that support the ceilings. Thus, the area beneath the ceilings is clear for audiences or activities. Perimeter-supported buildings are possible because the walls and the ceiling supports are of special design. The walls often have steel supports built in. The supports for ceilings may use lightweight materials. Design for these supports is geared to handle weights. The principles are much the same as those used for bridges.

Similar designs are needed in many factories. This is particularly true for buildings that house large assembly lines, such as factories that make automobiles. Large, clear areas are needed so that vehicles can be moved to a series of work stations. Specially supported ceilings and steel beams for the perimeter are common design techniques.

**Suspension bridges** are typically used for wide spans, as is the case with the famous Golden Gate Bridge, San Francisco. (Courtesy U.S. Department of Housing and Urban Development)

## BRIDGES

Bridges present a special structural challenge. People build bridges as crossings over rivers or canyons. Usually, bridges are built where there are no ready places to build foundations, or bases, for structures. Structural supports for bridges are built on shores or islands. Support points are minimal.

To bridge wide distances, special structure designs are used. Two common methods are:

- *Cantilever* construction uses a central tower as a support. The bridge is designed to spread its weight along the length of the tower. Usually, the parts of the bridge on each side of a tower are balanced. That is, the portions of a bridge on each side of a tower equal each other's weight. Structural design includes use of triangular and trapezoidal shapes within bridge supports.
- *Suspension* bridges hang from cables strung between support towers. Suspension means to hang. Large support towers are built first. Then heavy steel cables are strung between the towers. Roadway structures are then hung from the cables.

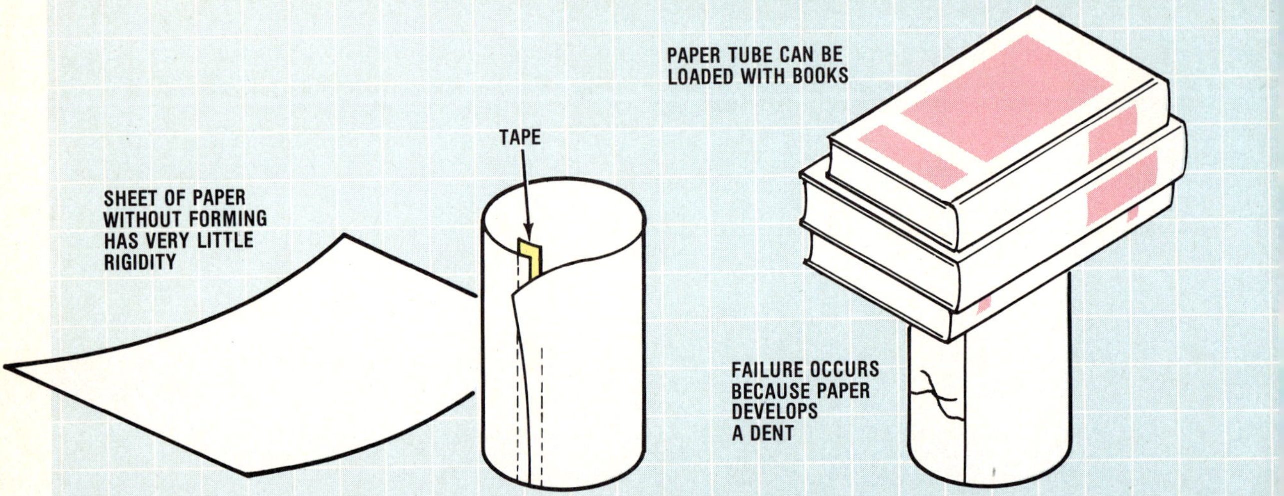


**This simple experiment** illustrates the strength that can be achieved through the use of shell structures. A simple tube collapses when weight is applied. However, if reinforced ribs are added, supporting strength increases greatly.

## SHELL STRUCTURES FOR VEHICLES

Different needs must be met in designing structures for vehicles. Bodies of vehicles such as automobiles or airplanes must be both strong and light. The light weight saves fuel because less power is needed to move the vehicle. At the same time, strength is necessary. Automobiles must be able to withstand bumpy roads and must protect passengers in case of accident. Aircraft bodies must be able to withstand high-speed winds and bumpy air currents.

A common technique of modern vehicle design is the *shell structure*. This is based on the principle that, with only light reinforcement, circular skins are extremely strong. You can prove this principle yourself by doing the simple experiment in the accompanying illustration.

Take a single sheet of typing or looseleaf paper. Form a circular tube by taping the ends of the paper together. Place a stack of books on the paper tube. Note that the books will fall as soon as a dent forms in the tube.

Now reinforce a paper tube by folding sheets of paper and taping four verticle strips inside a paper tube. When a stack of books is placed on the tube, dents will still form. But the reinforced tube will not crumble, and the books will not fall.

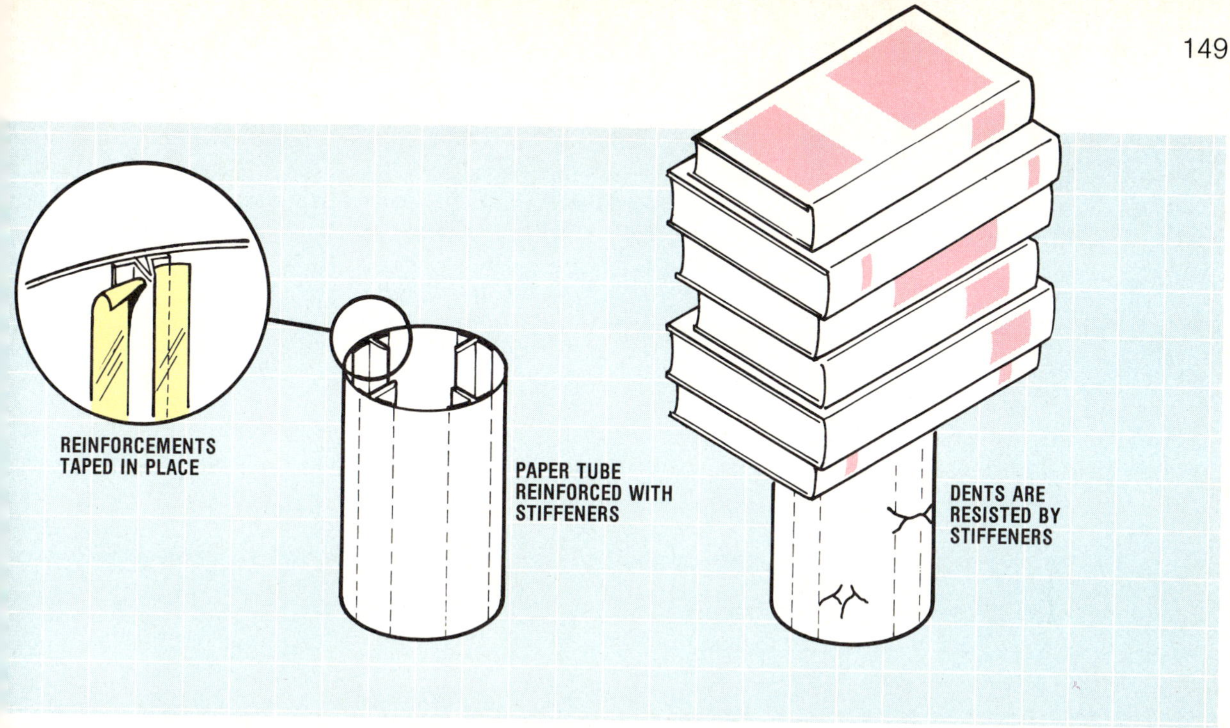


**Use of shell structures for modern cars** is illustrated by this exploded model that shows the different parts of the skin. (Courtesy Pontiac Division, General Motors Corporation)

**The building of shell bodies** has become a job for the modern technology of robotics. (Courtesy Cincinnati Milacron)

The same principle is used in the body, or *fuselage,* of many airplanes. A shell is constructed of a lightweight material such as aluminum. The shell is reinforced by interior structures called *bulkheads.* Also, rods called *stiffeners* and *stringers* are attached to the skin for extra support.

Similar design principles are used in the body of modern passenger cars. The body shell of a modern car uses the same basic principles as an airplane fuselage. Panels used to form the shell have metal ribs for stiffness and strength. Bulkheads are used for interior support.

## FORM AND FUNCTION IN STRUCTURES

A rule followed by most designers of modern structures is: "Form follows function." This means, simply, that structures are designed to do a job. The needs of the job are identified first. Then structures are designed for specific purposes. This principle applies to any type of structure, from modern homes to advanced space vehicles.

The same principle is also an important rule for any type of technology. The need comes first. Technology applies scientific principles to meet the need. The key word is "applies." Applying something is to use it in a practical way, to do a job. Technology applies science. In structural technology, then, the form of the structure meets the need of the function.

### Summary

People need shelter. Caves provided shelter for early people. However, people had to move frequently to find food, so they learned to build portable shelters from animal skins and poles. When people began to grow their food, they preferred permanent homes. The cumulative knowledge that people share about design and construction of buildings, bridges, and even vehicle bodies, makes up the technology of structures.

Commonly, structures are designed using basic shapes: Triangle, circle/arch, and square/rectangle. The triangle, probably first used in tent making, has three sides. People have found that combining triangles produces strong structures. However, the high and narrow structure is not ideal for comfortable living space.

Structures using circles and arches provide more living space than triangles and can be built without many tools. People wove branches or reeds into arched frameworks and used adobe or other natural materials to make hogans. Modern structures that use circular forms include bridges and hospitals.

Squares and rectangles did not become a common form in structures until people developed tools. Today, most rooms are either rectangular or square.

The framing supports a building, like the skeleton supports the body's organs and muscles. The exterior protects against the weather and provides structural support.

People developed new needs in structures with urbanization. Because land is scarce and expensive in a city, structures had to build up rather than out. Therefore, high-rise buildings were a welcomed solution. Skyscrapers could not be built, however, until the other technologies were developed. For one, frameworks that could support many floors were needed. For another, elevators to transport people and materials up to top floors were necessary.

Perimeter-supported structures are those that have strong exterior walls that support ceilings. This design requires no interior poles and, therefore, is ideal for entertainment centers or factories. The same principles of perimeter-supported structures can be applied to bridges.

Two common methods for constructing bridges are the cantilever and suspension bridges. Cantilever structures use a central tower for support while suspension bridges use two side towers.

Vehicle structures must be strong and light. Therefore, technology has produced a shell structure based on the principle that circular skins, lightly reinforced, are extremely strong.

The rule of structural design is that "Form follows function." This means that the needs of people using the structure determine how a structure will be designed.

## Technology Terms

story
foundation
triangle
triangular shape
framework
free-standing
base
pyramid
spire
gable
rafter
portable
permanent
flexible
circle
arch
radius
adobe

## Using Technology

1. Study the buildings on your school campus. How many basic shapes were used in designing these structures? What are they? Which shape was used more than others? Jot down your answers, either by drawing sketches of the structures or through written descriptions.

2. Often, people use common, available materials to build structures. For the Southwestern Indians, adobe was a common material. In parts of Arkansas, natural stone taken from local quarries is a common building material. What building materials are used often in your area? Can you discover why? Write your findings in a short composition.

3. Select one common building material—bricks, wood, etc.—and learn what you can about the material. What are the advantages and disadvantages of building with the material? How is the material made? What types of structures use the material often? Share your findings in class.

hogan
ellipse
square
rectangle
right angle
perpendicular
parallel
beam
framing
trapezoid
pentagon
high-rise
skyscraper
perimeter-supported
cantilever
suspension
shell structure
fuselage
bulkhead
stiffener
stringer

# III

# BIOTECHNOLOGY

An entirely different view of technology is presented in this part of your text. The four chapters in this part deal with the technologies of life. The first two chapters cover forms of life that provide food. The third and fourth chapters then cover the most important life form of all. This is the human body.

Chapter 8 describes the types of plant and animal life that people raise to feed themselves. This chapter describes the species of plants and animals that form the basis of agriculture. Agricultural technology is possibly the most advanced in the saving of human labor. Only some 3 percent of all workers are employed in agriculture in the United States. This relatively small group feeds hundreds of millions of people worldwide.

Chapter 9 goes on to describe the methods used to grow crops and deliver those foods for use by people.

Chapter 10 switches to the human body and to its life support systems. The vital systems of the body are described. The relationships among these vital systems are described. Also covered are methods for dealing with problems, or illnesses, that affect the human body.

Chapter 11 switches emphasis, discussing ways to maintain health rather than to care for sickness. Some basic body care measures—including diet and exercise—are covered.

# 8

# AGRITECHNOLOGY: PRODUCTS

## Your Learning Job

After you have read this chapter and completed the work assignments, you should be able to:

- ☐ Define agritechnology and explain how this system of technologies meets many needs and wants of people.
- ☐ List and describe five groups of plants that provide most of the edible food.
- ☐ Explain the advantages and disadvantages of a monoculture.
- ☐ Explain the differences between truck gardening and a monoculture.
- ☐ Describe how technology has been applied to the care and use of livestock.
- ☐ Describe how technology has changed the way farmers grow cotton.
- ☐ Describe how technology has changed the forestry industry.

Growing food is part of the job of agritechnology. Some 16 species of plants provide most of the food consumed by people. (USDA Photo)

## SCOPE OF AGRITECHNOLOGY

*Agriculture* is the science of farming. *Agritechnology*, or *Agritech* for short, is a system of technologies, involving many subtechnologies, that support agriculture. Agritechnology supplies many of the needs and wants that support human life.

Remember, the basic needs of people are food, clothing, and shelter. Most of these needs, in all three areas, are met by the products delivered by agritechnology:

- Virtually everything you eat is produced by agritechnology in the form of plant or animal products.
- The raw material for at least some of the clothing worn by most people is grown and produced by agritechnology. Agricultural products used to make clothing include cotton, wool, and leather.
- Agritechnology also grows the trees that provide the wood used in many homes and other buildings.

In addition to playing a role in meeting the needs of people, agritechnology also supplies many wants. For example, trees are the main source of paper. Paper is used to package food. It is also a tool for communication and, thus, is a tool for the accumulation of knowledge. In addition to providing clothing, agricultural products provide cloth for decoration, furniture, and some transportation items. For many centuries, for example, wood was the basic material used for building ships. Agritechnology provides materials for woven sails and ropes to control the sails. In these days of fuel shortages, agritechnology is providing alternative sources of fuel. Many agricultural products are being converted to fuels such as methanol, a substitute

**Many finished products for consumers** are produced from agritechnology materials. Included are wood and textiles. (Photo at left, Boise Cascade Corporation; at right, courtesy of Burlington Industries, Inc.)

for gasoline. In other words, agritechnology touches the lives of all people, every day.

Traditionally, occupations like farming are not thought of as being highly mechanized. However, in truth, agritechnology is probably the country's most mechanized occupation. Farming is also an extremely productive occupation. People who work in farming in the United States make up only 3 to 4 percent of the population. Yet, this small group feeds more than 200 million American citizens. American agritechnology also exports food that feeds a large part of the population of the rest of the world.

**Agricultural products can substitute** for some scarce resources. This facility converts surplus corn to gasahol, a mixed fuel that can be used to extend gasoline supplies. (Texaco Inc./American Petroleum Institute)

## Technological Emphasis

The physical technologies covered in the previous section of the book all apply to agritechnology. So, the use of tools and machines is not stressed in this chapter. Neither are physical technologies stressed in the remaining chapters on biotechnology in this section of the book. Instead, emphasis is placed on understanding the vital systems of living things (organisms). The important technologies in agriculture and health strive to understand living mechanisms. Therefore, stress is placed on descriptions of the makeup and functioning of plant, animal, and human life mechanisms.

**Agritechnology** has achieved extremely high rates of productivity. Some 3 or 4 percent of the work force is now involved in agriculture. These people feed the entire nation and much of the rest of the world. (Courtesy International Harvester)

## AGRITECHNOLOGY METHODS

As a system of technologies, agritechnology has many parts. These include:

- Land preparation
- *Cultivation,* or care and protection of growing crops
- *Irrigation*
- *Harvesting,* or the picking and collecting of food from growing plants
- Storing and handling of crops
- Food processing
- Breeding and development.

The technologies that make up the broad field of agriculture are many and can be complex. These technologies are described in the chapter that follows. In the remainder of this chapter, the discussions deal with agritechnology products—the actual crops that provide food, clothing, and shelter.

## AGRICULTURAL COMMODITIES (CROPS)

Farmers and scientists have been working for thousands of years to decide what plants and animals provide the best crops. Today, for example, there are some 700,000 known types, or *species,* of plants. Of these, some 80,000 are considered *edible.* An edible plant is one that can be used as food by animals or people. Of this group of plants, some 3,000 can be cultivated as crops to

**A variety of foods** is needed to keep people healthy and active. A balanced diet includes items from all of the major food groups.

produce food. The number of plants actually planted for crops is perhaps 300. However, among these, only 16 species provide some 90 per cent of the world's edible food. These 16 species are divided into five groups:

- Cereals are types of plants in which the seeds are the edible products. Five cereals are among the basic 16 species: rice, wheat, corn (maize), sorghum, and barley.
- Sugar plants are those that can be crushed and processed to produce a natural sweetener. Two sugar plants are included in the list of 16: cane and beet.
- Thrcc root crops are on the list of sixteen: potatoes, sweet potatoes, and cassavas (a type of sweet potato).
- Legumes are plants that have their seeds in pods. Three legumes are included: beans, soy beans, and peanuts.
- Three tree crops are listed: coconuts, bananas, and nuts.

All of the crops identified above are grown in quantity. All are important sources of food. However, this listing is just one of many that are available. Estimates of crop importance vary among authors and other sources. Some lists have only 12 crops; others have 16 or more. However, the point remains basic: Comparatively few edible crops feed all the animals and people in the world.

Some of the most important decisions being made about technology are in the field of agriculture. There is only so much land in the world. Further, only about 40 per cent of all of the land in the world can be used to grow crops. As the world's population increases, decisions about what crops to grow, on what land, become extremely important.

## Computers for Farmers

Large organizations and sophisticated technologies make these decisions about agricultural resources. For this reason, the field of agriculture has also become known as *agribusiness.* Technologies used by agribusinesses include sophisticated computer systems. Specific uses of computers include *modeling* techniques. A computer model represents a situation through use of mathematics. The computer is told about the type of soil to be used and the amount of land available. Expected weather conditions and other factors can also be included in models.

**Agritechnology uses satellites,** and many other advanced techniques, as research tools. Superimposed on the view of earth from space is a satellite image that analyzes the quality of water on earth. (Photo at left, NASA; at right, U.S. Department of the Interior/Bureau of Reclamation)

Managers of agribusiness companies then enter descriptions of choices for use of land. The computer tells them what results to expect for each choice. Management decisions are then based on computer-produced reports.

Another use of computers in agribusiness is to determine the diet for animals. Computers receive instructions about the size and age of animals being fed. Other information put into computers is about the food available for the animals. The foods are placed in large holding tanks in processing plants. Computers then mix the ingredients and deliver feed automatically. The computer programs determine how to give the animals maximum food value and growth for the least cost. These methods are known as *least cost feed blending.*

The discussions that follow identify the main agricultural crops. Note that many of these are products that you don't normally see on your dinner table. Some of the crops described below have to be processed before they appear in the main forms of food. In other cases, the crops are used to feed animals that produce meat, dairy products, fibers, and skins.

## Grains (Cereals)

*Grains* are products in which the seeds become the food items. Typically, grains grow on stalks or branches in clusters or on ears (corn). Few grains are used directly for food by people. Rice, for example, is considered a vegetable. Rice is packaged and cooked regularly for human use. Barley is used only occasionally in its natural form. Possibly the most popular use for barley is in soups.

Most grains are processed so that they are not eaten in their natural, growing form. For example, much wheat and corn is processed into *cereals* that are used as breakfast foods. Virtually all grains, particularly wheat, corn, and rye, are ground into flour and used for baking. Wheat flour is the most common ingredient for bread.

However, by far the greatest use for grains is as animal feed. This is particularly true for corn and sorghum. For example, you may have enjoyed eating corn on the cob, or ear. This variety of corn is known as sweet corn. Of all the corn grown, sweet corn represents an extremely small percentage. For every ear of corn eaten by people, thousands of ears are grown for animal feed. In producing animal feed, the entire corn plant is used. That is, whole fields of corn are cut and chopped up to feed to cattle or hogs. These varieties, taken together, are generally known as field corn.

The farming of grain is a major business. Consider that each of the grains identified above is grown in quantities of millions of tons. In some parts of the country, particularly in the Midwest, you can literally drive for many miles and encounter

**Most grain raised by farmers** is used for animal feed. Much corn, for example, is fed to hogs raised for pork and a number of by-products.

little but fields of corn, wheat, or other grains. A major source of United States income from other countries lies in the export of grains. These grains are used for animal feed and for flour around the world.

When large areas are devoted to the raising of single, specialized crops, it is called *monoculture.* In terms of technology, it can be efficient to have farming areas that specialize in individual crops. But there can also be problems. When land is used for a single crop, its natural resources—chemicals and minerals—tend to be depleted, or wear out. Eventually, fewer and smaller crops are produced on the same land. In parts of the South, for example, land was used entirely for growing cotton for perhaps a century or more. Eventually, that land became "farmed out." That is, the soil was so depleted that cotton could no longer be raised efficiently.

To avoid these bad effects of monoculture, special measures are usually taken. It is a common practice, today, to rotate fields

in areas practicing monoculture. For example, many wheat farmers divide their areas in half. They grow crops only every other year. In between, the land is fertilized and permitted to restore itself.

When monoculture is practiced, special attention must also be paid to the possibility of crop destruction by diseases and insects. For example, the boll weevil is an insect that attacks cotton plants primarily. In areas where thousands of acres of land are devoted to raising cotton, special care must be taken to look for the presence of the boll weevil. As soon as the insects are detected, special measures are taken to eliminate them. Similar problems exist for almost all species of crops.

The point is that, as with all technologies, agriculture has its own strengths and weaknesses. It can be very efficient to have crop specialization and monoculture. Raising and harvesting crops is very efficient when quantities of a crop are grown in a single area. However, concentration of crops also increases risks.

**Monoculture** is the concentrated growing of a specialized crop in a single area. Cultivating and harvesting such crops is efficient. However, special soil care may be needed. (Courtesy International Harvester)

**Quality must be monitored** continually to produce useful agricultural products and high yields. This photo shows a technical specialist examining rice plants. (USDA Photo)

Monoculture also provides opportunities for efficient uses of modern agricultural technologies. A number of specialized companies have brought high-production technologies to grain farming. Modern equipment for grain planting and harvesting is big and expensive. Few individual farmers can afford to own these machines. So, many specialized companies have been formed to operate the equipment. Farmers own the land and cultivate the crops. The work of planting and harvesting, however, is contracted to high-technology specialists. Employees of these planting and harvesting companies travel widely. They follow the planting and harvesting seasons. These specialists contract with farmers to plant grain. After the crops are grown, the trained crews return to harvest the grains. Specialized services of this type have contributed greatly to the productivity of modern farming.

**Experiments are conducted continuously** to determine the best use of land and the best feeds for animals. In this photo, a cow is being tagged with a number that will be used to keep track of an experimental program. (© Chuck O'Rear/West Light)

## Sugar Plants

*Sugar* is an important part of food. It is used in many of the baked goods or candies that you eat regularly. Also, sugar is commonly used in most processed foods packaged in cans or jars. Sugar is added to many vegetables during processing. Sugar is also present in soft drinks.

A problem with sugar is that you may enjoy it too much. Eating too much sugar makes people gain weight. Therefore, many people now limit the amount of sugar they eat. However, on a worldwide basis, sugar is still consumed at a rate of millions of tons per year.

Two widely different types of plants provide the source of natural sugar. One is sugar cane. Cane is, in effect, a giant grass with a central stalk. The other is the sugar beet, which is a root plant. That is, the beet grows underground, supported by a leafy plant that grows above the ground. Sugar beets are dug from the earth before processing.

With either type of plant, sugar is extracted by squeezing the basic plant—the stalk of the cane or the actual beets. The juices produced from the squeezing are then processed into sugar.

However, extracting the sugar is only the beginning of the useful life for sugar crops. Both cane and beets are valuable as animal feed. For many years, the processed cane or beets have been fed to cattle as a main form of nutrition.

More recently, processed cane and beets have taken on added importance as sources of fuel. These materials can now be processed to produce a form of alcohol known as *methanol.* Methanol can be combined with, or used as a substitute for, gasoline in internal combustion engines.

This final use of sugar plants helps illustrate an important feature of agritechnology. That is, agriculture produces renewable resources. These resources hold great potential as raw materials for new processing techniques. Consider: All of the oil in the world was once some form of living thing, including plants. Therefore, there is a relationship between natural fuels

**Sugar cane** grows well in warm, moist climates, such as Hawaii. (Courtesy Hawaii Vistors Bureau)

and agriproducts. Science has taken this challenge and is, continually, finding new and better ways to overcome shortages through agritechnology.

Use of agricultural products to create fuels shows how a technology, once developed, can be used in different ways. The technology that creates fuel from crops is *distilling*. Distilling techniques have been used for many years to produce alcohol. There are different types of alcohol. Some alcohols are used in industry and in medicine. Most alcohol, however, is used for beverages. Now, the same technology is being used to create methanol, which is conserving supplies of gasoline.

**Root crops** include plants, such as the carrot, in which the edible material grows beneath the ground.

## Root Crops

*Root crops* consist of plants in which the edible portions are the roots. Examples include potatoes, sweet potatoes, cassavas (similar to sweet potatoes), yams, carrots, and onions.

To get an idea of how root crops function, you can experiment with potatoes. Put a potato in a pan of water. Eventually, the "eyes" of the potato will sprout plants. This, basically, is the way potatoes are grown. Special, small potatoes, known as "seed potatoes," are planted. The new plants grow new potatoes as part of their root systems.

Some root crops, such as carrots, grow from seed. No matter what their source, all root crops present a common problem—difficulty in harvesting. Because the food products are underground, it is necessary to dig up the plants to get the product. Now, machines dig up potatoes and beets. However, there is still the problem of separating the plants from the roots that provide the food. Therefore, the harvesting of root crops requires extensive hand labor. Teams of migratory (traveling) workers move through areas where root crops are grown to provide this labor.

## Legumes

One characteristic of *legumes* is that the seeds of this plant are in pods. Examples include beans, green peas, soy beans, and peanuts. Another important characteristic of this plant lies in their chemistry. Legumes use the element nitrogen from air. Nitrogen appears in air as a gas, along with the oxygen that supports life when you breathe.

As a gas, the chemical nitrogen is *inert*. This means simply that nitrogen itself does not cause any chemical action or reaction. It is just there. Legumes, however, have bacteria in their root structures that use nitrogen from the air. The nitrogen is transformed to a water soluble chemical and trapped in the ground. The trapped nitrogen compounds are valuable fertilizers, helping plants to grow.

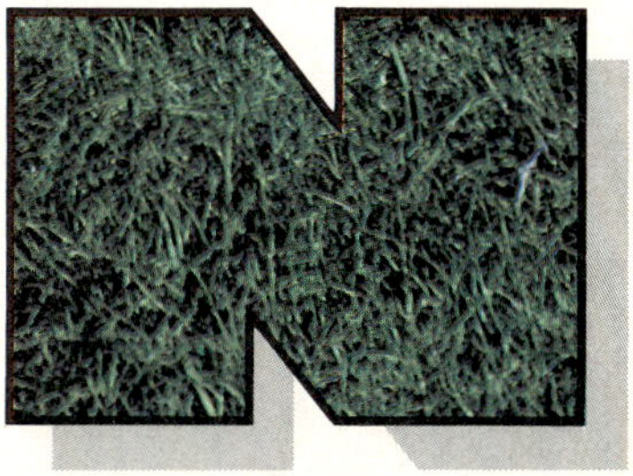

**Nitrogen is replaced** in the soil by legumes. The symbol N represents the supply of nitrogen provided by the alfalfa crop shown here. (USDA Photo)

Because of this nitrogen-forming property, legumes have special uses in agritechnology. Legumes are often used to help restore health to soil that has been "farmed out" on other products. This is especially true for one form of legume, alfalfa. Alfalfa is a valuable and nutritious component of animal food. In addition, when alfalfa is grown in a field, it fixes nitrogen in the soil, making that soil more fertile. Thus, it is a common practice to grow alfalfa in fields used for other crops on alternate years. The alfalfa can be sold for cattle feed. Farmers don't make as much money on alfalfa as they do on some other crops. However, the long-term value comes from rotating other crops with alfalfa to improve the condition of the land. The term *crop rotation* means simply that different crops are planted on the same land in different years.

Another advantage of alfalfa is that this plant provides relatively trouble-free crops. Alfalfa grows in great density, choking off weeds. Thus, alfalfa crops require little cultivation effort. Further, harvesting and processing are highly automated. Crops are cut and loaded into transport wagons in a single, rapid operation. Most alfalfa, today, is processed in automated plants. Intense heat is used to remove water content from the alfalfa. The dried product is then pressed into pellets used as cattle feed.

**An important legume** is the soy bean. This crop is valuable as a food source as well as for the replacement of nitrogen in the soil. (USDA Photo)

Many cattle growers produce their own alfalfa or contract for crops grown in fields of other farmers.

## Tree Crops

Certain edible foods, particularly fruits and nuts, grow on trees. Examples include the coconut, the banana, walnuts, pecans, almonds, and other nuts. Fruits that grow on trees include apples, peaches, apricots, pears, oranges, lemons, grapefruits, and avocados.

Farming areas used for tree crops are usually called *orchards*. Careful planning is needed to develop orchards. Enough space must be allowed between trees so that there is room for the branches to expand without crowding one another. Also, it must be remembered that fully grown trees cast shade. Therefore, the trees should not be allowed to cut off each other's sunshine. Trees are often trimmed to make harvesting easier. The taller the tree, the more difficult it can be to reach and pick the fruit. In addition, many trees must be cut back, or *pruned*, every year. This serves to shape the tree both for ease in harvesting and to maintain the health of the tree. For some types of trees, fruit grows more readily on new branches. Therefore, the old branches are cut away every year.

Traditionally, tree crops were picked by hand. Recently, however, a new technology has been developed for harvesting tree crops. Large machines literally take hold of the trees and

Examples of tree crops include apples, left, and walnuts. (Left, USDA Photo; right, Sun-Diamond Growers)

shake them until the fruits or nuts fall loose. These automated pickers have large, cushioned spreads to catch the fruits.

Most varieties of nuts used to present harvesting problems. These problems centered around removal of the nut meats from the shells. Large-scale users, such as bakers, couldn't afford to crack and shell the nuts by hand. In recent decades, the processing and shelling of nuts has been handled by a number of special machines.

## Vines

*Vines* are plants that grow long stems that cannot support their own weight. Therefore, as crops, vines must be supported on fences or supports that are provided especially for this purpose. The main vine-type crop plants are grapes, hops, and berries. Grapes and berries, of course, are eaten as fruits. They are also used to make jams, jellies, juices, and wines. The main use for hops is in making beer.

Usually, vines are cut back after each harvest to resemble small bushes without leaves. Then, each year, the vines grow anew and have to be strung on supports again. Therefore, considerable handwork is involved in growing vine-type crops.

The picking of grapes remains largely a hand operation. The packing and shipping of table grapes—those sold for use as a food—are also hand operations.

**Grapes are examples** of vine crops. Grapes produce juice and wine and can also be used as cattle feed. (Courtesy Wine Institute)

By contrast, the processing of grapes for juices or in wine-making is highly automated. Making quality wines is a slow process, however. This is because wines must age in large tanks, traditionally made of wood. The aging process can involve years of storage.

## Truck Garden Crops

*Truck gardening* gets its name from the traditional nature of this form of agriculture. Typically, there were farming areas around most major cities. The main markets were in the cities. Farmers would pick their crops and carry them to town in trucks.

In years past, there was a difference between crops that could be taken to market by truck and those that were shipped by rail. In those days, the railroads had refrigerated cars that could carry crops long distances. Until recent years, refrigerated trucks were unavailable. Today, however, either trucks or rail cars can be used for long-distance shipment. Vegetables and fruits grown on today's farms can be shipped all over the country. For example, lettuce grown in California is typically shipped by truck or train to the east coast.

Truck garden crops are also known as *row crops*. However, many products that are not connected with truck gardening are also grown in rows. Row crops include cotton, which does not fall in the area of truck farming.

**A row-crop example** is the pineapple. This photo shows the harvesting of pineapples in Hawaii. (USDA Photo)

For the most part, truck gardening involves the production of special fruits such as strawberries, as well as vegetables. Vegetables in this category include lettuce, tomatoes, celery, radishes, and cucumbers. Also included are cabbage, squash, broccoli, and melons.

As discussed in the next chapter, the ability to ship vegetables over long distances is, in itself, a major contribution of technology. People all over the country, in every type of climate, can now enjoy fresh vegetables all year round. Health problems used to develop in the winter because many people could not get fresh vegetables.

## Livestock

As described earlier, many crops are grown primarily as animal food. Providing food for animals has become a technology in itself. There was a time when animals were fed primarily on the natural plant life of the countryside. Dairy cattle raised for their milk were kept in pastures. They grazed on the grass that grew in the pastures and were called in twice a day for milking. Cattle grown for beef—and also sheep—were fed on ranges. These animals were simply turned out on wide areas of the countryside and permitted to eat whatever food they could find. Ranges were also called grasslands. In the western part of the United States, where there is little rain, grass can be scarce. At one time, it was said that it took 10 acres of Texas rangeland to feed one cow. (An acre is an area of land measuring 200 feet by 200 feet.)

Animals considered as livestock include cows, sheep, pigs, poultry, and goats. Cattle are raised both for beef and for dairy

**The barnyard** is the familiar home for many types of livestock including cows, hogs, and fowl.

products (milk). Sheep also produce two products—meat and wool.

Hogs (pigs) have characteristics that make them excellent sources of meat. This is because pigs use (digest) foods efficiently. A comparatively high yield of meat is delivered considering the amount of food consumed. Historically, pigs were kept on farms because they would eat almost anything, including the garbage left over by people. More recently, pig farming has become a high-technology business.

The same is true of poultry. Chickens and turkeys used to be kept in farmyards to eat whatever they could find, often consuming the same food as horses. Today, the poultry industry is highly efficient, producing considerable output in return for the grain consumed by chickens and turkeys.

Poultry farming also produces an important source of food in eggs. Actually, different breeds of chickens are used as the basis for meat- and egg-producing operations.

Goats are raised for the tasty cheeses that can be made from their milk. Some goats are also raised for meat and some for their hair.

**Livestock raising** involves the movement of animals to sources of food, including fields where they can graze. (Tourism British Columbia, Canada)

In addition, a major agritech business is building up around fish farming. Fish are raised in ponds for food in much the same way as animals are grown in pens. Like pigs, fish deliver high yields of meat for the food they consume.

Livestock management has become a prime area for the use of computers in agritechnology. As described earlier, most large operations now use computers to control the feeding of animals. In addition, livestock operators continually try to breed new, improved species. This applies equally to cattle, hogs, poultry, and even fish. Beef cattle are bred to produce large amounts of meat. The same is true for chickens and turkeys. Millions of information items must be recorded and tracked for scientific breeding. The processing of breeding records has been done with computers since the 1950s.

The dairy industry was a pioneering area for the use of technology in farming. Large-scale marketing of milk, for example, was made possible because of a scientific breakthrough. This was the process of pasteurization, named after the French scientist, Louis Pasteur. The problem was that disease-causing *bacteria* breed in milk. Bacteria are microscopic organisms. Some bacteria attack the bodies of people and animals, causing disease. Pasteur discovered a way to control the bacteria content of milk. This involves heating milk to kill the bacteria. Then the milk is cooled and kept cold as it is packed, shipped, and marketed. Special milk-processing plants now operate all over the country. In addition, the technology of refrigeration represents a multi-million-dollar part of the dairy industry.

**Milk** is an important product of livestock breeding. This old drawing of a milkmaid illustrates a tradition in farming. Modern farms use high-production machinery for milking.

Pasteurization purifies milk to a level that makes it satisfactory as a food. However, some bacteria remain after pasteurization. If refrigeration fails, or if the milk is permitted to stand for a long time, bacteria can multiply. Bacteria content of milk can reach dangerous levels. Also, the need for refrigeration has been a problem. It has been impossible to ship quantities of milk for children in parts of the world that lack sufficient refrigeration.

To serve people in these areas, a technology has been developed to produce *sterile* milk. Sterile means pure—free of disease-carrying germs. Sterile milk can be packed in air-tight containers and shipped without refrigeration. Milk can be stored without refrigeration for as much as a year or two. One method for sterilization of milk involves use of special machines. Milk is processed one drop at a time. The milk is brought to a high temperature that kills all bacteria. Then the milk is packed in sterile containers immediately—before cooling.

Another method is to pack the milk in airtight containers first. Then the packaged milk is sterilized through use of radiation. Exposing milk to x-rays or other sources of radiation kills bacteria without affecting the food products through which it passes. Radiation treatment can be used to sterilize a number of food products.

Another important technological development in the dairy industry has been the automatic milking machines. Milking cows was once a slow, time-consuming job that involved the efforts of many people. Now, large dairies have installed machines in clean rooms to handle milking automatically.

**Cotton** is one of the oldest agricultural crops in the world. This photo shows the harvesting of cotton. (USDA Photo)

## Cotton

Cotton is possibly one of the oldest agricultural crops in the world. Egyptians cultivated large fields of cotton some 5,000 years ago. Ideal climate and soil for growing cotton were found in the American South. In turn, the potential for farming cotton was one of the major reasons for colonizing the American continent.

A cotton plant grows to between two and three feet tall and has deep green leaves. The cotton itself develops in green pods that burst open when they are full of cotton. Within the cotton bolls (pods) are seeds that are rich in oil. This oil is used for many purposes. Cottonseed oil is used in food processing, for such products as margarine. In addition, the oil can be used in cooking and for finishing fine furniture. After the oil has been removed, cottonseed is used as an animal food. As important as cotton is, cottonseed is also a highly valuable agricultural product.

Cotton itself is classified according to the length of the fibers in the bolls. These fibers are spun during processing to form thread. Cotton thread, in turn, is used to make a widely used cloth. Many thousands of tons of cotton are grown in the United States each year.

The cotton industry has been affected dramatically by a major technology. Before either cotton or cottonseed can be

**Cotton grows** in bolls. When the crop is ready for harvesting, the white cotton bursts out of the green boll. Before use, the seeds must be separated from the cotton.

used, the seed must be separated from the boll. For centuries, this work was done by hand. In the early days of American independence, the separation of cotton from seeds was a major problem. There were simply not enough people available for this work. Cotton growing couldn't expand to its full potential without some solution to this problem.

The technology that solved this problem was developed by a young man named Eli Whitney. A year after graduating from college in 1792, Whitney was visiting on a friend's cotton plantation in Georgia. On the plantation, he learned of the problem of separating seeds from cotton. Whitney responded by inventing a machine to do this job. The new device was called the *cotton gin.* The word gin was slang in the Georgia area for engine. The first model of a cotton gin was produced in 1793. A patent was granted in 1794. Almost immediately, the acreage used for growing cotton in the South multiplied.

The term gin is still used to describe modern cotton processing facilities. Whole wagonloads of cotton are now processed continuously. Cotton is packaged in bales (wrapped bundles) of approximately 500 pounds (230 kilos) for shipment. Seed is separated for processing of oil and feed.

Areas of the South have depleted their lands after centuries of cotton farming. Because of another technology, however, cotton farming has expanded and has moved to the southwestern and western states. The new technology is *irrigation.* Irrigation delivers water to crops in areas that don't have enough rain for natural farming. Areas such as West Texas, Arizona, and Central California have ideal climates and land for cotton. However, these areas lack natural supplies of water. With irrigation, cotton has become a major crop in these areas. The technology of irrigation is discussed further in the chapter that follows.

## Forestry

Trees grown for their wood represent a special type of agritech product. Most agricultural products are produced on an annual basis. That is, there is a new crop each year. Even for livestock, animals raised for meat are marketed in time spans that run from six weeks to a maximum of two years. Tree farming is different. It takes at least 15, sometimes as much as 25 years, to grow a tree.

**The technology of forestry** involves the replacing of trees that have been cut down. New trees are grown in hothouses, then planted to create forests. (Both photos, USDA/Forest Service)

Tree farming is relatively new in the United States. At one time, the entire continent was rich in forests. People simply cut trees as they needed wood or fuel. Also, many trees were cut down and burned just to clear the land they grew on. Forests were turned into farms. As supplies of trees grew scarce, renewal programs developed. Today, it is common practice to replant young trees wherever old trees are cut down.

The main end products of forestry are wood and paper. To get some idea of these, consider papermaking. It takes some 17 full-grown trees to make a ton of paper. A single newspaper in a large city might use 20 to 40 tons of paper daily.

Because it is an agritech area, forestry has some special characteristics. Trees grow slowly. However, trees are used in large quantities. Therefore, it takes many, many square miles of forests to support this branch of agritechnology.

A major technological development that has affected forestry is the *chain saw*. The key to using forest products, of course, is to cut down trees. For centuries, this work was done with axes or with long saws operated by two persons. One logger stood at each end of the saw. The saw was passed back and forth. This method took many hours to cut down a large tree. Today, a single person with a chain saw can do the job in minutes. As a result, costs of logging have come down. Production in the forestry industry has grown.

## Summary

Agritechnology is a system of technologies that applies the principles of agriculture, mechanics, and even computer science. The products of agritechnology supply food, clothing, and shelter for people. In addition, paper, furniture, and even fuel can be made from agricultural products. Agritechnology is one of the most mechanized occupations in the United States.

The system of agritechnology includes land preparation, cultivation, irrigation, harvesting, storing, processing, and development. One subtechnology that helps this vast system is computer technology. Computers help farmers and business people to make decisions about crops and land use.

The country's food supply is provided by five basic groups of edible plants: cereals, sugar plants, root crops, legumes, and tree crops. Grain can be ground to flour for bread or fed to animals. Today, most grain is farmed in a monoculture system, which devotes land to a single crop. A monoculture is an efficient method of farming, but it can lead to problems. Land can be "farmed out," or entire crops can be destroyed by a single disease or insect.

Sugar plants, in addition to being processed for sweeteners, are used for some animal food as well as methanol, a fuel. Root crops include potatoes, carrots, and onions. Legumes are plants that help restore nitrogen to "farmed out" soil. Thus, legumes, particularly alfalfa, are used in crop rotation plans.

Tree crops provide many fruits and nuts and are usually grown in orchards. Vines also provide food products, such as berries and grapes. Truck garden crops, or row crops, refer to food products that are carried to consumers by trucks or trains. Most vegetables and many fruits are row crops.

Livestock management has become an important part of agritechnology. Today, most animals are fed carefully mixed portions of grain or other food products. Even the fishing industry has changed. Now, people have learned how to "farm" fish. Technology has been applied to the processing of animal products. For example, agritechnology has developed a way to ship sterile milk without refrigeration.

One product of agritechnology—cotton—has been cultivated to make cloth for thousands of years. Now, agritechnology has found a way to grow cotton in dry climates—by irrigation. Another area of agritechnology—forestry—has also changed considerably. Now, forests are farmed. As soon as one tree is cut down, another tree is planted.

In short, agritechnology seeks to use land as efficiently as possible. The decisions made in this area of technology will affect all people.

## Using Technology

1. Peanuts offer an example of a legume that is used is a variety of ways. Name several food and nonfood products that use peanuts. Besides the numerous ways in which peanuts can be used, what is another advantage in growing this plant? Write your answers in a short composition.

2. Farmer Sam pays $1.00 for each quart of picked peppers. Howard picked six pecks of peppers. With the money he earned, he bought a guitar. The price of the guitar was $45.00. How much money did he have left? (Hint: Your dictionary will have a weights and measures table.)

3. It takes 17 trees to make one ton of paper. Suppose that your town's newspaper business uses 15 tons of paper daily. The paper is issued every day—seven days a week. How many trees would be needed to supply the paper for one week?

## Technology Terms

agriculture
agritechnology
cultivation
irrigation
harvesting
species
edible
agribusiness
modeling
least cost feed blending
grain
cereal
monoculture
sugar
methanol
distilling
root crop
legume
inert
crop rotation
orchard
prune
vine
truck garden
row crop
bacteria
cotton gin
irrigation
chain saw

# 9

# AGRITECHNOLOGY: METHODS

## Your Learning Job

After you have read this chapter and completed the work assignments, you should be able to:

- ☐ Describe how land is prepared for farming.
- ☐ Explain different methods for planting seeds.
- ☐ Name several techniques for cultivation of crops.
- ☐ Explain three ways to irrigate crops.
- ☐ Describe the problems of harvesting crops and how technology tries to solve these problems.
- ☐ Explain how technology has affected food processing in recent years.
- ☐ Illustrate how breeding or development can improve food production.

**Agriculture** has become a high-technology business with the introduction of powerful tractors and high-capacity cultivating equipment like that shown in this photo. (Courtesy International Harvester)

## THE TECHNOLOGIES OF AGRICULTURE

A field as vast as agriculture consists of a system of technologies. The basic technologies of agriculture are identified in the previous chapter. The previous chapter also discusses these methods in a limited way, as they apply to the products of agriculture. In this chapter, the methods themselves are reviewed. These methods are:

- Land preparation
- Planting
- Cultivation
- Irrigation
- Harvesting
- Storage and handling
- Processing
- Breeding and development.

## LAND PREPARATION

The purpose of farming is to use land effectively. Remember, there is only so much land in the world. Further, only about 40 percent of that land is usable for farming. Therefore, planning how to use suitable land so as to produce as much food as possible is important. The objective of productivity is to get as much food as possible from each unit of available land.

Productive use of land starts with preparation. There is no really natural farm land. The very idea of farming involves a commitment to prepare the land and to use it in special ways. The first step is to clear land of its natural growth and other

**Land preparation** involves breaking up the soil so that new plants can grow freely. For most types of farming, land should also be flattened so that water flow reaches all of the plants in a field. (USDA Photo)

obstacles to farming. One of the historic requirements in preparing farm land is to cut trees growing naturally on the land. Then, farmers have to remove the stumps and roots of these trees.

During the pioneering days in the United States, land clearing was a major job. Many people earned fortunes by clearing land, preparing it for farming, then selling it. People would move into the wilderness where land was available at low cost or no cost at all. They would clear the trees and prepare the land for planting. They might grow one or two crops. However, because the land was cleared, it was worth a lot more than its original cost. People sold the land and moved on to another location where land was cheap or free.

In addition to trees, many areas of the country required clearing of rocks before land could be used. In some sections, particularly in parts of New England, stones removed from fields were used to build fences around them. Stones were

gathered on flat, wooden platforms called stone boats. The stones were then drawn by horses to locations where they could be stacked into fences.

For some types of crops, the land must be almost completely flat before it can be used. One crop which requires flat, level land is rice. In parts of California, new technologies use laser beams to survey, or measure, the land to be sure it is flat before planting starts. A *laser* is a concentrated, high-power beam of light. A laser is called *coherent light* because of the way its rays are formed in tight patterns.

In hilly or mountainous parts of the world, land preparation involves *terracing*. Terracing is a major activity in land preparation in Japan, where there is a shortage of land for farming. Terracing, basically, is a technique to create flat areas for farming from hills or mountains. Flat, step-like sections are cut into hillsides. These are known as *terraces*.

**Terracing** creates flat areas usable for farming or housing. This technique is used in hilly areas. (© Craig Aurness/West Light)

Terracing is not used extensively for farming in the United States. However, at present, terracing is used to develop locations for building homes. This is done particularly in western areas where cities are built in hilly or mountainous locations.

After land has been cleared and leveled as necessary, *plowing* takes place. Plowing is also called *tilling* the soil. The purpose of plowing, or tilling, is to dig up, or turn the soil. In small gardens, the soil is sometimes turned with a shovel, or spade. The shovel is simply stuck into the earth, and the dirt is picked up and turned over. A plow increases productivity. A plow, simply, is a shaped steel device that is stuck into the earth. As the plow is pulled, the earth in its path is turned. The row of earth turned by each plow is called a *furrow.* In early days, plows were pulled by animals. For modern, mass-production farming, equipment is available that has many plows on a single, tractor-drawn unit.

The purpose of plowing is to loosen the soil. Remember, high-production farming cannot be done on natural land. When land is plowed, the soil is loosened so that the roots of the crops can take hold and spread.

As part of the plowing operation, *fertilizer* is sometimes added to the soil. Fertilizers are natural compounds or chemicals that make the soil fertile. *Fertile* means that the soil contains

**Contour plowing** traps water for distribution to crops when land being farmed is not completely level. (USDA Photo)

elements that encourage plants to grow. Fertile soil feeds, or nourishes, plants.

Often, fields are plowed in patterns. These patterns depend upon the characteristics of the crops and the land. Different types of crops require different distances between furrows, or rows. For example, cotton requires considerable activity in weed control while it is growing. Either people or machines must get in between the rows of cotton to eliminate the weeds that could use up the plants' nutrition. On the other hand, corn grows high and creates its own shade. Comparatively few weeds grow after the corn reaches a midpoint in its development. Therefore, rows of corn can be placed closer together than rows of cotton.

Special plowing techniques are needed if the land has slopes or hills. At one time, it was considered almost impossible to use hillsides for crop farming. This is because water naturally runs downhill. Therefore, the high areas in a sloped field would be dry while the lower areas would have too much water. Recently, *contour plowing* techniques have made slopes available for productive farming. In contour plowing, rows are not drawn in straight lines. Instead, the furrows follow the shape, or contour, of the land. The rows are drawn across the slopes so that they trap water rather than letting it run off. This is an example of how technology can make more resources available and use those resources more productively.

**Fertilizing or seeding** of farmland is done automatically with tractors after land has been prepared. (USDA Photo)

## PLANTING

*Planting* is the placement of seeds into ground that has been prepared to receive them. Seeds, of course, grow into the plants that produce food.

Methods of planting depend on the type of crop to be grown. For most crops that grow in rows, *drilling* techniques are used for planting seed. That is, a rod or tube is actually pushed into the ground. Seeds are then dropped through the tubes and placed beneath the surface of the earth. Examples include corn and cotton, crops for which plants are grown in rows, at specific distances between plants.

Another technique that is commonly used is *broadcast seeding*. Under this method, seeds are simply scattered over the surface of the ground. The ground is prepared as a flat surface prior to seeding. In some techniques, the ground is raked or turned lightly to fix the seeds. In some instances, the earth is watered after seeding.

If you have ever watched someone plant a lawn, you have seen broadcast seeding. Special spreaders are used—or the seed is spread on the ground by hand. Then, the ground is usually raked over to keep the seeds trapped in place. After that, the ground is usually watered thoroughly. Similar techniques are used for planting many types of grains or grasses. Examples include wheat, alfalfa, and rye.

When broadcast seeding is used for crops to be planted over a wide area, airplanes or helicopters are sometimes used to scatter the seed.

## CULTIVATION

*Cultivation* is the care of growing plants. Cultivation takes place after the time when seeds are planted and continues until crops are harvested. The main purpose of cultivation is to control weeds and insects. Weeds are plants other than the crops to be harvested. Any other type of plant growing in a field uses nutrition that the crops need. Thus, other varieties of plants are eliminated. Similarly, insects and plant diseases can destroy crops. They too must be controlled.

The age-old method of cultivating fields is to use a *hoe*. This is a small tool designed to turn the earth. Hoeing digs up and covers weeds. Today, especially on large farms, cultivating is

**Cultivating** controls weeds and potential germ infections of plants while they are growing. (USDA Photo)

done by machines. Large cultivators can turn the earth and control weeds in more than a dozen rows at a time.

Other techniques for weed control include *herbicides* and *mulching*. Herbicides, simply, are chemicals that kill plants. Herbicides have been developed for specific crops. They kill any plant life in the field other than the crop being grown. Mulching is a technique for spreading materials around the plants and between the rows where they grow. These materials, in effect, strangle the weeds before they can grow. One mulching technique is to spread straw around the roots of plants. Another is to use strips of plastic. The crops are planted through the plastic. The plants grow through and above the plastic sheeting. The

**Airplanes are often used** for crop dusting that achieves pest control while plants are growing. Crop dusters can be either biplanes or single-winged monoplanes like the one shown. (USDA Photo)

plastic then traps moisture in the soil to feed the plants and also eliminates most weeds. Crops grown with plastic covers include strawberries and pineapples.

Insect control is achieved by spraying plants with either liquids or powders. For large fields, much of this work is done by *crop dusters.* Crop dusters are either airplanes or helicopters that spray or dust fields in which crops are growing. The work of some crop dusters has been dramatized. These are specialized pilots who fly biplanes at low altitudes and slow speeds. Biplanes are aircraft with two wings. Biplanes are usually used because the extra wing surface provides lift to enable the planes to fly relatively slowly. Some biplanes, for example, can fly at speeds as low as 40 miles per hour. Usually, crop dusters fly 10 feet or less above the height of the plants they are dusting. Sometimes they even fly below power lines crossing the countryside. These specialized airplanes provide a quick, inexpensive technology for controlling insects in large farming areas.

Another method for controlling insects that can destroy plants is through natural means. So-called friendly insects are turned loose in the fields. These insects are enemies of the ones that can destroy the crops. The good insects, in effect, eat the bad ones.

## IRRIGATION

Water is necessary for the growing of crops. This water must be fresh; that is, it may not contain any salt. The most traditional and widely used source of water is rain. In an area that receives enough natural rainfall to support crops, irrigation is unnecessary. However, many areas with fertile land and ideal growing climates do not get enough rain to support crops. In these areas, technology took over to deliver the water. In many areas of the

West and Southwest, agriculture is the main user of available water. For example, in California, more than 80 percent of all the available water is consumed by agriculture.

Three major methods are used to irrigate crops:

- Gravity Flow
- Sprinkling
- Drip.

**Gravity flow irrigation** permits water to flow naturally through planted fields. (U.S. Department of the Interior/Bureau of Reclamation)

## Gravity Flow

*Gravity flow* uses the natural attraction of the earth for everything in the atmosphere. Water naturally flows downhill. This is in response to the attractions of gravity. When you drop something or throw something up, it eventually falls to earth. In irrigation technology, gravity can provide the force for delivering water to crops.

To use gravity flow, a supply of water must be at a level higher than the planted fields. Gravity flow can be natural. This occurs when water is taken from a river or stream and directed into fields. Some gravity flow systems use normal, hilly countryside to control the flow of water. Water is held in artificial lakes, or reservoirs. Dams hold the water and create the lakes. From these reservoirs, water flows in canals to areas where it is needed for drinking or irrigation.

As an alternate method, *ground water* is placed in a special tank. Ground water is stored, usually, in natural reservoirs under the ground. The supply of ground water in any area is called the *water table.* Wells or pipes can be sunk into the underground water table. Power can then be applied to pump the water out of the ground. This power can come from electric motors or from windmills. In any case, the water is pumped into a tank that is stored above the ground. When the water is released from the tank, gravity flow takes over.

In a typical gravity flow system, farmers can open a gate or small dam. This allows water to flow in ditches through the fields. From the irrigation ditches, the water just flows through the rows between the furrows of the fields. Plants are irrigated by this flowing water.

## Sprinkling

Most *sprinkler systems* are controlled by some form of power. Water is pumped into a *spray* device. A spray is a device that forces water through small openings, or holes. A spray that you

**Sprinkling** irrigates crops in a manner similar to natural rain. (USDA Photo)

are probably familiar with is a shower head. In irrigation, the effect of spraying is to create a kind of artificial rain. Sometimes long pipes are used. Other times, powerful sprays cover hundreds of feet of a field at one time. The sprays are moved as the fields are irrigated.

## Drip

*Drip irrigation* uses mechanisms that release water drop by drop. Think of the faucet in a sink. If you turn it on just slightly, water will come out a drop at a time.

A drip irrigation system delivers water through a series of tubes with valves that can be controlled like faucets. The valves can be set to control the amount of water that is delivered to each point in the system. Typically, a drip system has many valves that deliver water to individual plants.

A drip irrigation system produces the most growing power from the least amount of water. Thus, the drip technology has been popular in areas that are short of water supplies.

In addition, the crops grown have to have relatively high values. This is because drip systems are relatively expensive.

In addition to being valuable where water is scarce, drip systems have other uses. In particular, drip systems are ideal for crops that are grown on hillsides. Water can be pumped to the top of the hill, then allowed to drip through the system by gravity. With the availability of drip systems, hillside areas that could not previously be used for farming are now very valuable. An example is the growing of avocados in California. Avocado trees are often planted on the hillsides. There they receive large

**Drip irrigation** uses valves at points where water is actually to be delivered. Thus, scarce water supplies are limited to individual plants or areas. The photo at left shows a drip irrigation system. At right, an outlet delivers water to a single, small area. (USDA Photos)

amounts of sunlight and heat, ingredients they need for growing. Water is dripped through a series of tubes to all of the plants, individually. Before drip irrigation was available, avocados could not be grown on hillsides in this way.

Drip irrigation has also become popular for vineyards planted on hillsides.

## HARVESTING

*Harvesting* involves picking, cutting, and/or gathering crops. Harvesting has a special problem. The work, when it is done by hand, is a slow process that requires many people. When crops are planted and cultivated, farms do not need large numbers of people. This requirement comes just at harvest time. For this reason, harvesting has been done, for many years, by *migratory labor*. Migratory laborers, or migrants, are people who travel from area to area picking crops. Migratory means to move from place to place. Birds, fish, and animals all migrate to climates and food supplies. People migrate to sources of employment.

Migratory labor has been a problem for the United States. Since these people move to areas for short times, there is usually a shortage of comfortable places for them to live. In some areas, migrants have been housed in tents, without running water or toilets. Whole families often move with the crops. This means

**Harvesting techniques** are adapted to the needs of individual crops. For example, the harvesting of pineapples, top, left, is done with conveyors taken out to the fields to carry the hand-cut fruit. Grapes, right, are too delicate for machine picking. Grapes are hand picked and deposited in carts. By contrast, the harvesting of flax is a machine process. (Top left photo, courtesy Hawaii Visitors Bureau. Right photo, courtesy Christian Brothers. Bottom, left, USDA Photo)

that children do not have a normal school experience. There have been situations in which children of migratory laborers had no opportunity to go to school at all.

Health care can also be a problem, since doctors in any given area may be overworked. At best, these doctors will not know about their patients or have records of their health histories.

Another problem is that, as wages for migratory laborers have increased, costs of food have risen. This has caused hardships for people who consume the food. One of the reasons for this is that harvesting is a particularly expensive operation for many crops. When harvesting is done by hand, and when people are paid full wages for their work, the price of crops increases. In some cases, this has meant that foods grown in the United States have lost consumer sales volumes. This has happened because the same foods grown in other countries sell for less.

Foreign-grown foods have price advantages because workers are paid less than workers in the United States. American farmers lose income under such conditions.

Because of the importance of harvesting costs, technology has been applied increasingly to this job. The previous chapter describes some methods for applying technology to harvesting. For example, rather than picking crops out of trees, there are now devices that shake trees and catch the nuts or fruits as they fall. There are also many automatic machines that harvest crops.

One of the best examples of how technology has advanced in harvesting can be seen in the case of wheat. Wheat is a grain used to make flour for bread and cake. Wheat is grown in large fields. Individual farms may have thousands of acres of wheat.

The process of harvesting wheat involves cutting the entire plant, or stalk. Then the wheat is shaken, or beaten, to release the seeds. The pods that surround the seeds on the plant are

**Shaking things up** is the way walnuts are harvested, left. Special devices shake the trees and catch the nuts as they fall. Photo top, right, shows the harvesting of rice. Wheat is being harvested in the photo at the bottom, right. (Photo at left, courtesy Sun-Diamond Growers. Top right photo, courtesy Free China Information and Communications Division. Lower right photo, courtesy Sperry-New Holland.)

**The traditional harvesting tool** was the scythe. When tools like this were used, 80 percent or more of the work force was involved in agriculture.

disposed of separately. These are known as *chaff.* In addition, the stalks of the wheat plant themselves must be disposed of. This material is called *straw.*

A major invention in the history of agritechnology has been a machine that could harvest wheat. The machine cuts and gathers wheat directly from the fields. This machine was invented by Cyrus McCormick and patented in 1834. It was known as the McCormick *reaper.* In agriculture, the word *reap* means to cut and bring together crops.

Reaping is just the first step in the harvesting of wheat. The next operation is *threshing.* This involves removing the wheat seeds, known as *kernels,* from the plants. Once the reaper had been accepted, McCormick invented a thresher. Together, the reaper and thresher made it possible to harvest wheat entirely by machine.

Later, the earlier machines were improved with the development of *combines.* Combines do both reaping and threshing in a single, continuous operation. The first combines were introduced in the 1920s. Today, wheat is harvested by specialists. People with equipment that costs hundreds of thousands of dollars travel over wide areas of the country, harvesting crops under contract from farmers. The modern machines package the wheat in bags and also package the chaff and straw for further use. A modern combine can do a complete harvesting job at 12 acres of wheat per hour.

Similar uses of technology have been applied to many crop areas. The development of technology for harvesting is still going on.

## STORAGE AND HANDLING

Moving and storing farm crops or food materials is a big business. As just one example, the United States government spends $1 million each day to store foods. These are foods that have been bought from farmers or food processors under government programs. The government then holds these materials for future use or for sale when they are needed.

Thousands of railroad freight cars and trucks are used to haul farm produce each day. This is a basic part of agritechnology. That is, foods are grown on farms. Farms, necessarily,

**Shipments of grain** originate from grain elevators of this type. Crops are delivered to elevators following the harvest. The ducts seen in the photo are then used to move the grain into railroad cars for shipment. (© Craig Aurness/ West Light)

are in areas where not many people live. Thus, foods grown in the country are eaten mainly in the city. Transportation is essential.

Further, harvests take place at specific times of the year. The results of the harvest, however, may have to last an entire year, or even more. Thus, it is necessary to have massive, special buildings in which to store farm products. These buildings are designed specifically for use of individual types of products. For example, grain products are usually stored in *grain elevators*.

These are usually round, tall structures. They get their name from the fact that grain is loaded into the tops of these storage buildings. Usually, grain is carried automatically, on conveyors, from railroad cars or trucks to the necessary level and dropped inside. Grain is then taken out at the bottom, through gravity feed. The United States has capacities for storing millions of bushels of grain. A bushel contains 32 quarts of material.

For some products, such as milk, butter, and meat, *cold storage* is necessary. Large refrigerated buildings hold millions of tons of cold products. To get an idea of what is involved, think of the size of a typical refrigerator at home and the cost of running it. Now consider the cold storage of products throughout the United States. This job involves a series of buildings that could occupy entire city blocks. Placed side by side, these cold storage buildings would occupy as much space as a large city. It is not hard to see how the government's food storage bill runs to $1 million dollars daily.

Food storage costs and facilities operated by private companies are even greater than those maintained by the government. Remember, all of the farm products harvested at any given time cannot be processed immediately. So massive buildings are needed to hold products while they are waiting to be processed. Then, after processing, other buildings hold the products until they are sold.

## PROCESSING

*Food processing* is possibly one of the world's most mechanized industries. Processing takes in everything from receiving the crops to cooking and packaging foods. Typical processing facilities are food canneries. Truckloads of food are cooked and flavored at one time in large cooking vats. Then, these prepared foods are carried away on automatic conveyors. The foods are delivered to machines that fill and close cans or jars at rates of thousands per hour. These machines also line up the cans or jars and place them in boxes, ready for shipment.

Flour milling is virtually all automatic. Wheat flows through a processing stream that includes grinding, screening to take out large particles, and packaging.

In recent years, the frozen food industry, which had existed for some time, has developed major new technologies. In the

early days of frozen food, the foods were prepared normally, then stored in large refrigerators. It could take days before the foods were frozen enough to be packaged for marketing.

More recently, *cryogenic* methods have been introduced. Cryogenic temperatures are those that occur when gases, such as hydrogen or oxygen, are placed under great pressure. Under pressure, the gases become liquids. Cryogenic temperatures are in the range of 320 degrees below zero, Fahrenheit. In cryogenic processing, food is said to be *flash frozen.* That is, freezing takes place in a matter of seconds or minutes. Foods are frozen, ready for delivery, as they move down conveyor belts. Quality is better because the flavor of the food is captured more completely. Also, processing is quicker.

For packaging some food products, nuclear energy can be used productively. Nuclear radiation is a form of energy that can

**Storage and handling** of agricultural products is a massive job. Grain elevators of this type store wheat, corn, or soy beans. (© Craig Aurness/West Light)

**Food processing** has become highly automatic, as shown in this facility that crushes sugar cane and extracts the juice. (Courtesy Sugar Cane Institute)

pass through solid materials or liquids. The radiation has the effect of sterilizing the foods, killing all germs. Once sterilized, foods can be kept indefinitely, without refrigeration.

Radiation treatment of foods has been useful in killing germs for flour to be used in bread. As described in the previous chapter, radiation is used also for sterilizing milk. Sterilized milk can then be stored without refrigeration.

In recent years, there have been great increases in automatic techniques for meat processing. At one time, whole sides of meat—each representing half an animal—were delivered to butcher shops or supermarkets. The meat then had to be cut up and processed by hand in individual stores. More recently, this work has been centralized in large processing plants. Stores receive cuts of meat that are ready to sell.

The meat industry is recognized as one of the most efficient in the use of its raw materials. Virtually everything that comes into a meat processing plant leaves as a salable product. The meat is sold for food. Hides are used for leather. Even hair is used for special purposes. Some animal hair is used for strings on musical instruments. Other types of hair are used for bristles on brushes. The guts, or intestines, of some animals are used to produce strings for tennis rackets. Gut is used also for stitching materials used by surgeons. Anything that is left over is used either as animal feed or for fertilizer. It was once said of a plant that processed hogs that it used "everything but the pig's squeal."

## BREEDING AND DEVELOPMENT

Agritechnology deals with living things. Originally, people took their food the way they found it. Animals, grains, fruits, and vegetables were consumed just as they grew in nature. Often, this meant that early people had to wander over great distances to find supplies of food. You have undoubtedly heard stories about how American Indians wandered the plains in search of buffalo for food.

Gradually, people began to process food in the same way as they made clothing or other manufactured items. That is, specialization took over. Entire fields or even areas were assigned to grow single crops. As described in the previous chapter, this is known as monoculture.

**The raising of animals** can be an interesting and profitable pastime for young people in rural areas. (Texas Department of Highways/ Travel and Information Division)

Once specialization set in, invention was not far behind. People studied the methods of breeding, or reproduction, of plants and animals. Farmers found that controls could be applied to develop foods more efficiently and in greater quantities. In effect, people learned how to put nature to work for them. The processes involved are natural. That is, nature itself develops plants and animals individually. Take a large field of corn, for example. No two plants will be exactly the same. Some plants will have more corn and larger ears of corn than others.

People studied these developments. The seeds of the largest plants in a crop were saved for the next planting. Experience showed that the next crop would grow bigger plants that produced more corn. The same principle held for animals. The eggs from the larger chickens were saved. These eggs were then used to produce chickens that, in turn, laid bigger eggs. Chickens and other animals were also selected for characteristics. Poultry were selected on the basis of larger breasts because there is more meat there. Cattle were selected for their properties to produce milk or meat.

**New varieties** of plants are developed continuously. In this facility, seedlings are planted and grown as part of a continuing series of breeding experiments. (Courtesy Sugar Cane Institute)

Through the centuries, plants and animals have been selected and bred to produce more and better food. The original selection and breeding steps can be called a science. This science is known as genetics, or the study of heredity—the passing of traits from one generation to another. However, when the scale of modern agriculture is introduced, breeding and development also become technologies. Picking and experimenting with the seeds from individual plants of corn can be considered genetics. However, processing entire fields of corn and producing seeds for next year's crop is a technology.

Considerable work and frustration can be involved in the breeding and development of foods. In one recent example, scientists worked hard to develop a tomato that would be easier to pick and to pack. Tomatoes were selected and bred for larger sizes and thicker skins. The idea was that these fruits would not bruise if they were picked by automatic machines. Another feature was that all of the tomatoes on a plant ripened at the same time. This meant all the tomatoes could be picked at once. After considerable numbers of experiments, scientists were able to develop tomatoes with exactly these properties. There was only one problem: The new fruits did not taste like tomatoes and, therefore, could not be sold as tomatoes.

Everybody has to eat to stay alive. As the world's population has grown to more than 4 billion people, food production has become a major challenge. Agriculture has become a primary area of technology. Work in agritechnology has been interesting and productive so far. The challenges continue. Agritechnology still has many breakthroughs to make. The field continues to be interesting and important, both for consumers and for agritech professionals.

## Summary

Agritechnology is a system of technologies, all of which involve agriculture. Technology has developed farming methods to use land effectively. That is, the purpose of many technological advances is to produce more products on less land.

First, land must be prepared. If forests exist, the trees must be cut down. If the land in stony, the rocks must be cleared. Then, the soil must be tilled, or dug up.

Next, seeds are planted. Methods for planting vary according to the type of plant. Row crops are usually planted by drilling techniques. Broadcast seeding is a method that scatters seeds over the surface of the land. Most grasses are planted in this manner.

Once the seeds begin to grow, the plants must be cultivated. Cultivation controls weeds and insects. One way to cultivate crops is by hoeing.

Another way to kill weeds is to use herbicides. Still another method is mulching. Generally, insects are controlled by spraying fields. Large fields in some areas are sprayed by crop dusters. Some destructive insects can be controlled by other insects.

Water is a necessary element of plant growth. In areas that do not receive enough rain, fields must be irrigated. The three major irrigation methods are gravity flow, sprinkling, and drip. To use gravity flow, there must be a water supply higher than the fields. Then, water is released and flows naturally to fields. A sprinkler system requires power. Water is pumped into spray devices to create artificial rain. Drip irrigation delivers water much like a faucet. The slow supply of water makes this method especially efficient for dry areas.

Once the plants have matured, the crops are harvested, or gathered. One problem that has affected farmers and consumers alike is the cost of labor in harvesting crops. One important aim of technology has been to develop machines to harvest crops. Harvested produce must be stored or shipped to consumers. The technology of transportation is important in this part of agritechnology. Many products must be refrigerated or stored until processing can take place.

Food processing involves receiving and packaging food. Technology has changed the way food is processed. For example, food can be sterilized by radiation. The meat industry has developed ways for using virtually every part of animals. What is not used by people is made into animal feed.

Agritechnology also involves breeding and development of better plants and animals. People have learned to use seeds from their best plants for future crops. The same methods can be applied to animals. Breeding a cow that produces more milk is one method to increase efficiency in agriculture.

### Technology Terms

- laser
- coherent light
- terracing
- terraces
- plowing
- tilling
- furrow
- fertilizer
- fertile
- contour plowing
- planting
- drilling
- broadcast seeding
- cultivation
- hoe

## Using Technology

1. Suppose that you were given 640 acres of farmland. You are to grow one of the following crops: cotton, fruit trees, or corn. Before you decide, what must you know about where your land is located? What will you need to know about the farm equipment in the area? What are the special problems of each crop? What are the advantages of each crop? Write your answers in a composition.

2. If one acre of land yields 20 bushels of wheat, how many bushels will three square miles produce?

3. Suppose that one type of tomato, Red and Ripe, yields 25 tomatoes per plant. However, producers have discovered a problem: The tomatoes bruise easily. Producers report that up to one quarter of their crop is lost when the tomatoes are shipped. Another type of tomato, Red Sweet, yields 15 tomatoes per plant. Producers note that Red Sweet tomatoes are hardy during shipment—few tomatoes are lost. Which tomato would you choose to grow and why?

herbicide
mulching
crop duster
gravity flow
ground water
water table
sprinkler system
spray
drip irrigation
harvest
migratory labor
chaff
straw
reaper
reap
threshing
kernel
combine
grain elevator
cold storage
food processing
cryogenic
flash frozen

# 10 HEALTH CARE

## Your Learning Job

After you have read this chapter and completed the work assignments, you should be able to:

- ☐ Explain how health care benefits all people.
- ☐ List and describe the basic systems of the human body.
- ☐ Explain how people can read vital signs to determine the condition of a human body.
- ☐ Trace the life cycle of a human in terms of health.
- ☐ Describe several threats to human health.
- ☐ Describe the way that pharmaceutical technology responds to health problems.
- ☐ Define diagnosis and describe three categories of diagnostic technique.
- ☐ Describe methods used to treat ill people.

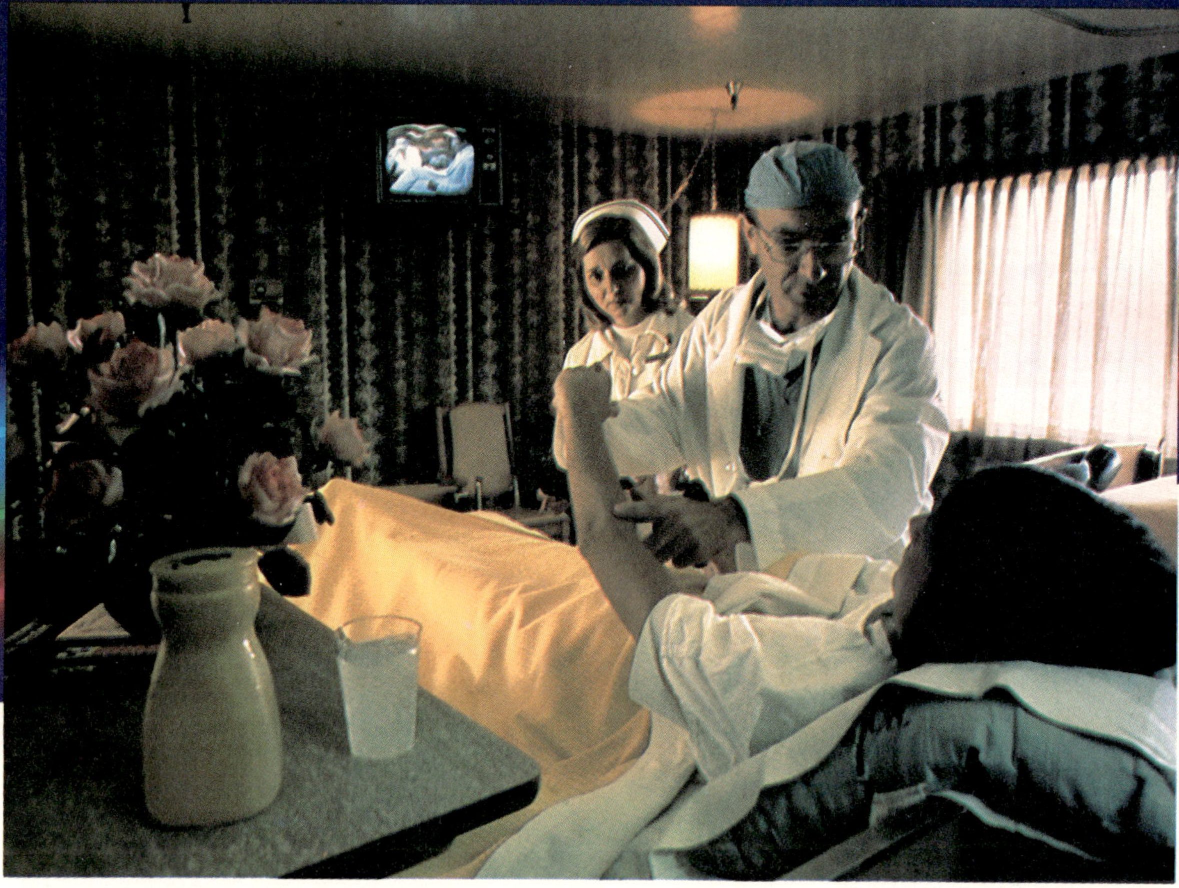

Health care concentrates upon the treatment and cure of illness. (Courtesy AMI)

## HEALTH AND ITS CARE

One of the world's vital resources in achieving progress through technology is the human body. People are thinkers and builders. Without the efforts of people, there would be no technology, no society as it is known today. Therefore, the care of the human body becomes a vital technology in itself.

*Health care* is a term that takes in a number of technologies. The term itself has two key meanings. First, *health* is the word that describes the state or condition of the human body. A healthy body is one in which all of the parts, or systems, are working as they should. Therefore, the term health tends to mean good health, or sound health.

When something is wrong with a human body, its condition is usually described in terms of illness, or sickness. Different types or degrees of sickness refer to different levels of loss of health. If someone is slightly, or temporarily, sick, the condition can be described as a mild illness. People might say that person

**Good health shows.** Attention to the principles of health care can help people feel better and live longer. (Florida Division of Tourism)

is "under the weather." If conditions are worse, people can be described as seriously ill or even critically ill. Persons who have illnesses that will lead to death are said to be terminally ill.

The objective of health care technologies is to keep people well, or healthy. Healthy people, in turn, have active minds and bodies. Healthy people invent and build. Healthy people benefit other human beings.

The result of good health care, in turn, is longer human life. At the beginning of this book, it is mentioned that in prehistoric times, a 14- or 15-year-old person would have been middle-aged. In those days, the average person did not live much past the age of 30. As people studied health and developed technologies for health care, people lived longer. That is, people had a greater *life expectancy,* or *life span.*

The relationships between health care and life span can be seen even in present-day situations. The life expectancy of people differs widely in different parts of the world. For example, in much of Africa, the life expectancy for the average person is still 42 years. By comparison, in the United States and Canada, life expectancy is 73 years.

In Africa, there are more than 9,000 persons for each physician, or doctor. In the United States, there is one physician for each 600 persons. The United States has an average of one hospital bed for each 133 persons. In Africa, the figure is 550. Clearly, there is a relationship between the quality of health care that people get and the length of their lives. Since life is the most important possession anyone has, health care is viewed as a vital technology.

## TECHNOLOGY AND LIFE

The human body itself can be viewed as a group of systems. That is, each person has a group of life-support systems within his or her body. The body, in turn, depends upon the normal, continuing function of these systems to maintain its health.

Detailed study of the human body is part of the science of biology. This book on technology does not go into detail about the body and its systems. However, the technologies of health care are all designed to study these systems and keep them functioning normally. Therefore, to learn about the technologies of health care, you should have an understanding about the body

**Life expectancy** can increase if proper health care is available. People now live longer than they did in the past. (UNRWA)

systems that give a person health. The basic systems of the human body include:

- The *circulatory system* distributes blood throughout the body. A central pump, your heart, drives the blood supply in a continuous loop. The heart pumps blood through a series of *vessels.* Blood vessels carry blood to different parts of your body. Then, another set of vessels carries the blood back to the heart. The vessels that carry blood from the heart to the body parts are called *arteries.* The vessels that return blood to the heart are called *veins.* This full circle of blood from the heart to the body parts and back is called *circulation.* This is why the entire system is called the circulatory system.

  As blood travels through the body, it delivers the necessities of continued life—food, fluids, and oxygen. In addition, the blood returning to the heart carries waste materials the body will dispose of. Waste materials are routed through special organs that deal with these functions. Part of the circulation lies in picking up food and oxygen from some organs while dropping off waste materials

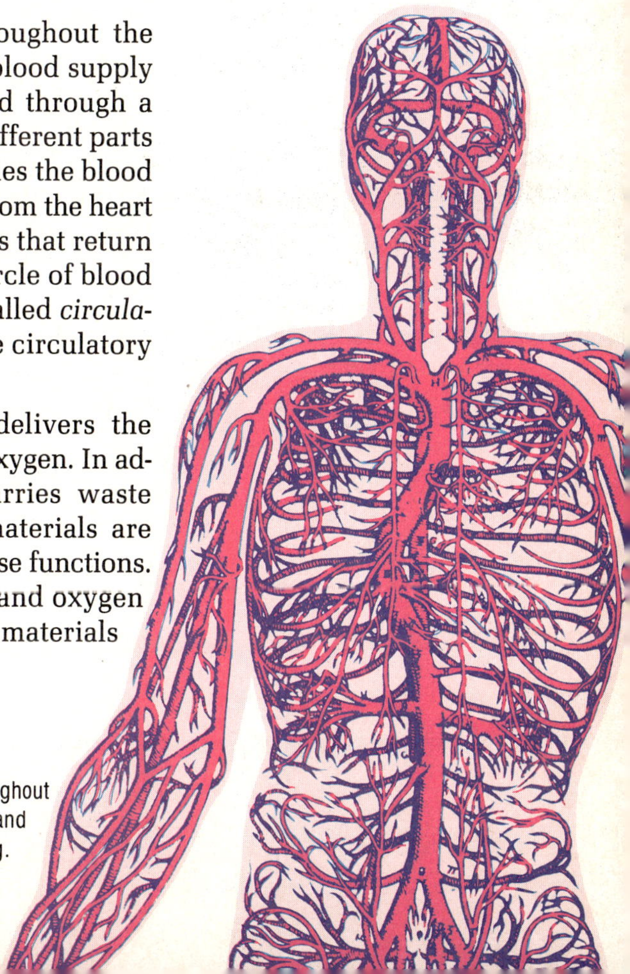

**The circulatory system** delivers blood throughout the body. Blood, in turn, is the source of fuel and oxygen that the body burns to keep functioning.

at others. The full cycle of feeding and disposal is handled each time the blood circulates through your body. In a normal body, your heart pumps blood anywhere from 70 to perhaps 140 times per minute. The rate at which the heart pumps depends on what you are doing. If you are exercising, the heart pumps more rapidly.

**A strong respiratory system** is needed for many activities. The respiratory system provides the oxygen needed so that the body can burn fuel. Many musicians need well-developed respiratory systems. (The Salt Lake Valley Convention and Visitors Bureau)

- The *respiratory system* delivers oxygen to circulating blood. Oxygen, as you know, is necessary for combustion. Your body is in a continuous state of combustion. That is, you burn fuel to produce energy. Fuel is the food you eat. Oxygen makes possible the combustion of this fuel, or its use as energy. *Respiration* means breathing. When you breathe, you draw air from the atmosphere around you. The air contains oxygen. Mechanisms within your lungs draw the oxygen from the air and feed it into the blood stream. At the same time, the lungs remove the gases that result from the burning of fuel within your body. Thus, the lungs, or respiratory system, take in oxygen and put out gases.

  The main gas discharged by the respiratory system is *carbon dioxide.* This is a chemical produced by the process of burning. Carbon dioxide, for example, is also created by the burning of such fuels as coal or gasoline. Note that the respiratory and circulatory systems are related. They depend upon each other. The heart pumps blood in a double cycle. One cycle carries the blood from the heart to the lungs and back again. In the process, the lungs remove carbon dioxide from the blood and insert oxygen. Your lungs then breathe out the carbon dioxide that your body has to eliminate. The oxygen picked up by the blood in this first cycle is then returned to the heart. In the other cycle, oxygen is delivered to the remainder of your body.

- The *nervous system* makes it possible for the body to feel and to react to external conditions. The *nerves* that reach from your brain to every part of your body are like electric or telephone wires. Nerves receive *stimuli,* or messages, about the conditions faced by the body. When the body touches something hot, the nerves deliver one kind of message. If your body is cold, a different message is delivered to the brain. The brain is the hub of the nervous system. It is as though your body has its own telephone network. The brain

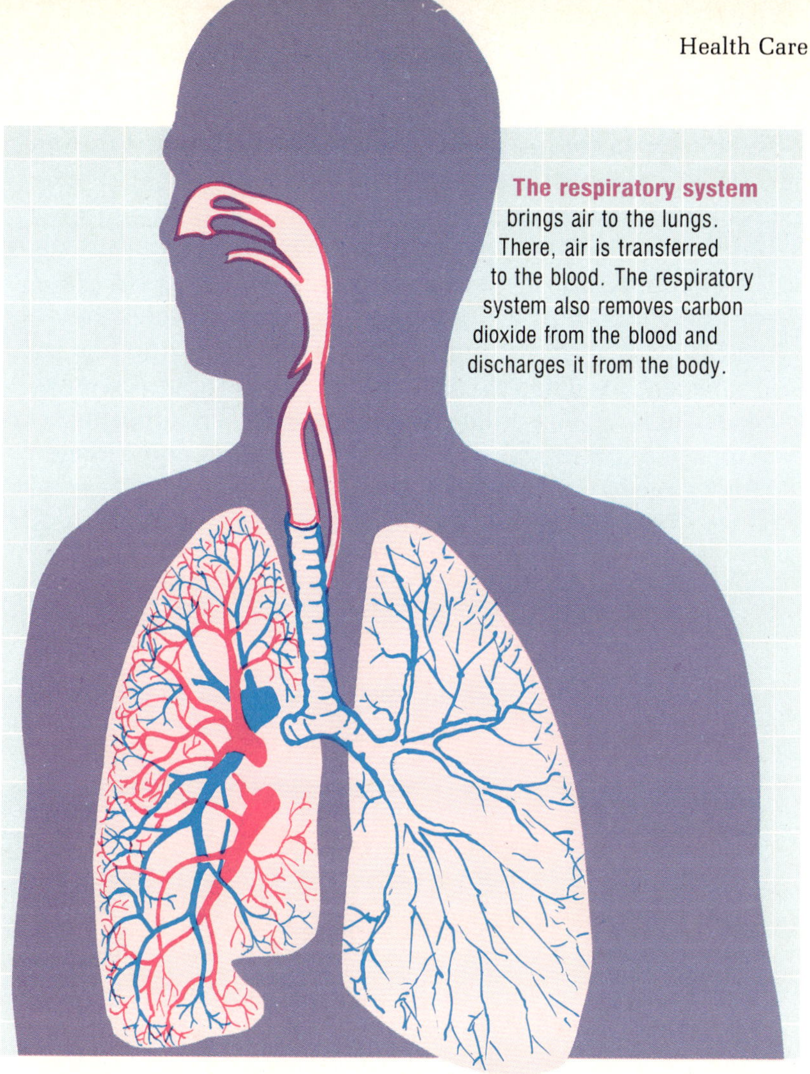

**The respiratory system** brings air to the lungs. There, air is transferred to the blood. The respiratory system also removes carbon dioxide from the blood and discharges it from the body.

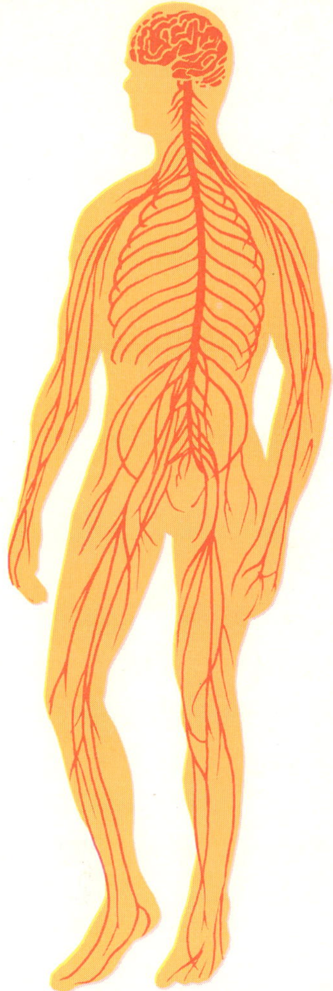

**The nervous system** makes it possible for the body to sense and react to external conditions.

is the central office, where everything is connected. Your fingers, arms, legs, and other parts of the body are the subscribers, or customers.

Anything you feel, hear, smell, taste, or see becomes an *impulse*. An impulse is a signal carried by the nervous system. The impulses tell the brain what is happening to the body. The brain, in turn, uses the nervous system to send out signals. These signals tell the body parts what to do. If you touch something hot, the nervous system causes you to pull your hand away. Suppose you are waiting for the start of a race and you hear a gun sound. The brain tells you to start running. If you see words on paper, as in this book, the nervous system creates impulses. The brain reacts to these impulses, storing the information you are seeing.

- The *muscular-skeletal system* gives the body its structure and its ability to move. The *skeleton* is the set of bones that supports the body. Muscles are special tissues that cause the body to move. Muscles cause movement by *contracting*. Contraction is the act of becoming shorter or smaller. When you use your muscles, they shorten, causing movement. Your body has sets of muscles. When you shorten one set of muscles, for example, this causes your arm or fingers to bend. Then, when you relax those muscles, an opposing set of muscles contracts, causing the first set to become longer. This causes your arm or fingers to straighten. Muscles and bones work together to give your body form and movement.

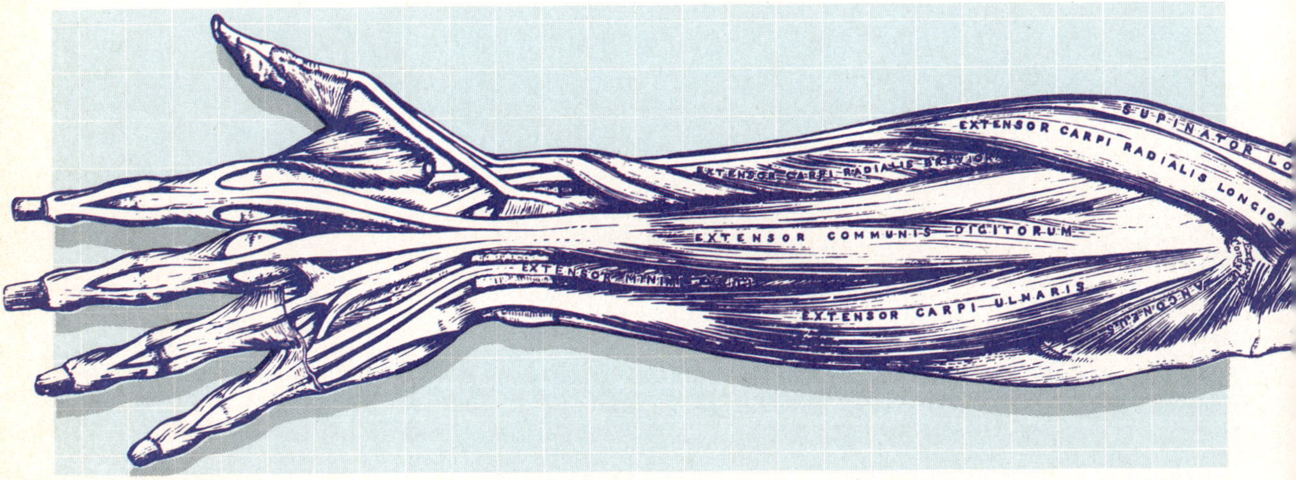


**The muscular-skeletal system** gives the body its structure, strength, and ability to move.

- The *digestive system* converts the food you take into your body into usable fuel. Digestion is performed in your stomach and intestines. Your stomach receives the food you eat and passes it along to the intestines. In the intestines, processing is completed so that tho chemicals needed for food are put into the blood stream.

- The *excretory system* gets rid of, or eliminates, the bodily wastes. *Excretion* means to eliminate, or get rid of. Part of the excretory process of the body has already been described. That is, your lungs excrete carbon dioxide when you breath out. Vital excretory functions are also served by the kidneys. In a sense, the kidneys separate the blood chemically, then reunite it in a purified form. In the process, waste products of combustion, particularly *nitrogen,* are removed from the blood. These nitrogen compounds are then carried away from the kidneys as liquid wastes that your body eliminates. In addition, the excretory system also carries out the solid wastes left over at the conclusion of the digestive processes.

- The *reproductive system* carries on life from one set of people, or generation, to another. Reproduction involves all of the processes needed for the development and birth of infants. Reproduction is part of the life cycle. People are born. They develop and grow. In the process, they give birth to children who follow the same cycle. Thus, reproduction is essential for continuing life. The reproductive systems of males and females are different. The female system includes the organs needed to carry, or hold, new bodies while they are developing. The female system also feeds babies until the time they are born. The female system also has the ability to provide nourishment to infants after birth. The male system, in turn, provides part of the stimulus that causes the reproductive system to start. Both male and female reproductive systems contribute traits that form the new organism. That is, each child contains elements of the traits of both parents.

**The digestive system** enables the body to convert food into usable fuel. Thus, the makeup of a person's diet can affect the body's ability to digest and use food.

- The *endocrine system* reacts to and stimulates all the other systems. *Endocrines* are chemicals called *hormones.* Hormones, in turn, have been called chemical messengers. They

**At times of great activity** the body's endocrine system stimulates all of the other systems. (Indiana Department of Commerce)

are carried through the blood and affect the functions of the main organs of the body. One of the hormones that your body uses is *adrenaline.* When you get excited, your body pumps extra adrenaline. You become more alert. You can run faster or exert extra strength.

The human body, then, represents a kind of technology of its own. There are a group of related systems that function together to control life and development. The normal functioning of the body systems, in turn, controls health. When everything is normal, you are healthy.

## LIFE (VITAL) SIGNS

The body systems are like technologies in their ability to display information about their own condition. On an automobile, for example, simple tests of condition can be applied. There are gauges that tell you fuel level, temperature of coolant, oil pressure, and other key operating indicators. The body has similar capabilities. Trained people can read these *life signs,* or *vital signs,* of the body. These key signs include:

- Pulse
- Blood pressure
- Respiratory rate
- Temperature.

### Pulse

The *pulse* is a reading of the rate at which your heart beats. The pulse rate can be measured at any *artery.* An artery is a blood vessel that carries blood away from the heart. Blood surges, or spurts, as the heart pumps. These surges, or spurts, are counted to indicate the rate of operation of the heart.

As a life sign, the pulse rate has different meanings under different conditions. At one extreme, the at-rest pulse rate for a healthy, active person can be as low as 35 or 40 beats per minute. For a normal person, a usual pulse rate is 65 or 70 beats per minute. The pulse rate changes dramatically as a person exercises. A person who is running or walking quickly can have a pulse rate ranging from 125 to 160 beats per minute.

The most usual point for measuring pulse rate is the inside of the wrist at the base of the thumb. The pulse rate can be taken on either hand. Use the fingers of the other hand to feel the pulse. Use a wrist watch with a second indicator to count time.

**The body's pulse rate** increases or decreases according to need. With exercise, the pulse rate increases. When the body is at rest, the pulse rate decreases. (Florida Division of Tourism)

Count pulse beats for either 10 or 15 seconds. If you count beats for 10 seconds, multiply by six to get your rate for a full minute. If you count for 15 seconds, multiply by four.

It is a good idea to check your pulse rate from time to time. Take readings both when you are at rest and at pauses in exercising.

## Blood Pressure

*Blood pressure* measures the force exerted by the blood as it moves through arteries. Blood pressure is measured through a technology that uses pressure. An area of the body, usually the arm, is wrapped tightly. The tightness is usually caused by pumping air into a cuff that is wrapped around the arm. Filling the cuff with air creates pressure on your arm. This creates a means of measuring the pressure applied by the blood flowing through your arteries.

Blood measurements use electronics technology today. Sensitive devices measure and display the values of the blood pressure at two points. One point is at the peak of the pulse, or surge, of blood. The other is at the low point, when only a minimum amount of pressure is being applied. Before electronic technology was available, instruments used mercury, a heavy fluid, to measure blood pressure.

## Respiratory Rate

The *respiratory rate* is the number of times per minute that you breathe in and out. At times when you are exercising, your body needs more oxygen. You should have a higher respiratory rate at such times. When your body is resting, your respiratory rate should come down. At rest, typical healthy bodies have respiratory rates of 15 or 20 per minute. Under the strain of exercise, the respiratory rate can go as high as 35 or 40.

Doctors often use a special instrument to listen to your body and to monitor its life signs. This is known as a *stethoscope*. The stethoscope contains a flat, round device that is placed against parts of your body. This is shaped to pick up the sounds of your heart, lungs, or pulse. These sounds are transmitted through tubes connected to earpieces. A trained doctor or technician uses a stethoscope as a special tool to monitor some of your vital signs.

## Temperature

To function, the body uses fuel. This means that some form of combustion is taking place within your body. Combustion, as you know, produces heat. A normal, healthy body, therefore, should have a normal, healthy body temperature.

The average temperature of a human body is 98.6 degrees F (Fahrenheit). On the Celsius scale, the body temperature is 37 degrees. These temperatures are considered average, or normal. Remember that everybody is individual. There may be slight variations in normal temperatures for different people. Also, body temperatures may vary for the same person. For example, your own body temperature probably varies through the average day. Your temperature tends to be relatively low when you first wake up in the morning. Later in the day, body temperature normally increases. This is because your body temperature rises as combustion rates increase, and as your body disposes of more wastes.

In checking the condition of a body, a sign to watch for is a higher-than-normal temperature. If your temperature rises, this could mean that your body is fighting some illness. To fight the illness, you are burning additional fuel. Your body becomes warmer.

Body temperature is measured with an instrument called a *thermometer.* A thermometer measures heat. Traditional thermometers have a small amount of mercury in a sealed glass rod. The mercury expands up a scale to indicate temperature. Electronic thermometers are also used. These provide direct readings, in numbers, to indicate temperatures.

Together, the life, or vital, signs are indicators of your body's health. When the vital signs are normal, your body's systems are working as they should—individually and together.

**Exercise strains** the body's systems. Exercise can be good for the body's vital organs. (© John Patsch/Nawrocki Stock Photo)

## LIFE CYCLE

Each body goes through stages. At its present stage, your body is either going through or preparing for major changes. These changes involve growth and development. The sound of your voice and the shape of your body may be changing, or will change soon. This is part of a normal cycle. All human bodies go through a cycle that affects the need for health care.

From the time a baby is born, change and development take place continuously. For the first one to three years of life, a baby is building the basic strength to survive. Newborn babies need help in eating, in holding their heads up, and in turning over when they are lying down. As a baby gains strength, health also improves. However, during the first years of life, there are great dangers to health. There are also major needs for health care and health technologies.

One of the measures for the quality of health care is the *infant mortality rate.* This is the measure of the number of deaths to babies under one year of age for each 1,000 births. Today, the infant mortality rate in the United States is 14. In Africa, the rate is 147, or 10 times as great. For children between the ages of one and four, the average death rate in Africa is 35.6. In the United States, the rate is 0.8. These figures demonstrate how health care

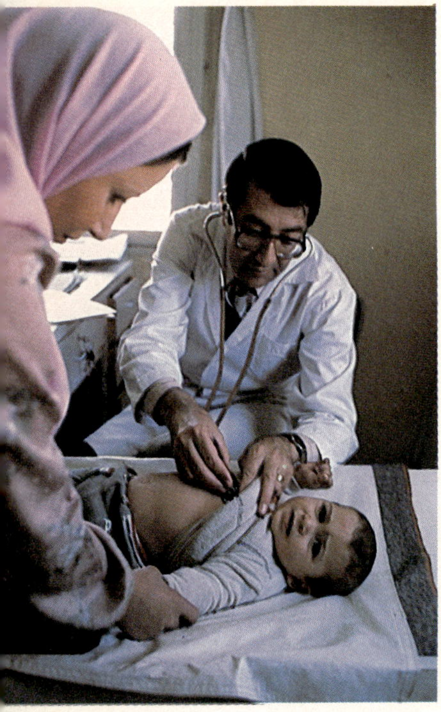

**Health care requirements** are at a high point in the early years of a person's life. (UNRWA)

technology supports life. In addition, the same figures show something about the life cycle. Once an infant survives the first year, the chances for continued life improve greatly. Between the ages of one and four, the death rate is just a fraction of what it is during the first year of life.

In general, people develop and enjoy relatively good health during the years when they are growing up and functioning as young adults. During these years, the leading cause of death is accidents. Barring accidents, people are usually pretty healthy.

During years of growth and young adulthood, time works for people. As individuals grow older, however, time can become an enemy. The body begins to decline after a person passes his or her peak years. Starting in a person's late thirties and forties, parts of the body simply begin to wear out. As age increases, physical strength decreases. With age also, the need for health care increases. As shown in the accompanying diagram, health care needs are high in the beginning and end of life. In between, most people are relatively strong and healthy. This is basically the cycle for most normal human bodies.

## THREATS TO HEALTH

The term health care is not entirely accurate. Actually, health care is needed primarily by people who are sick or injured. In this sense, health care is a *treatment* for persons who are in less than normal health. As described above, some health care is needed because young bodies are developing and older bodies are declining in health. In addition, some part of the population needs the services of health care technologies at any given time. This care deals with a number of causes for loss of health. These causes can include:

- Diseases
- Trauma
- Environmental problems
- Psychological problems
- Genetic conditions.

### Diseases

Throughout the world, and throughout time, living organisms have consumed (eaten or used up) one another. For food, people use plants and animals. It also works the other way. There are many smaller organisms that exist at the expense of humans. For

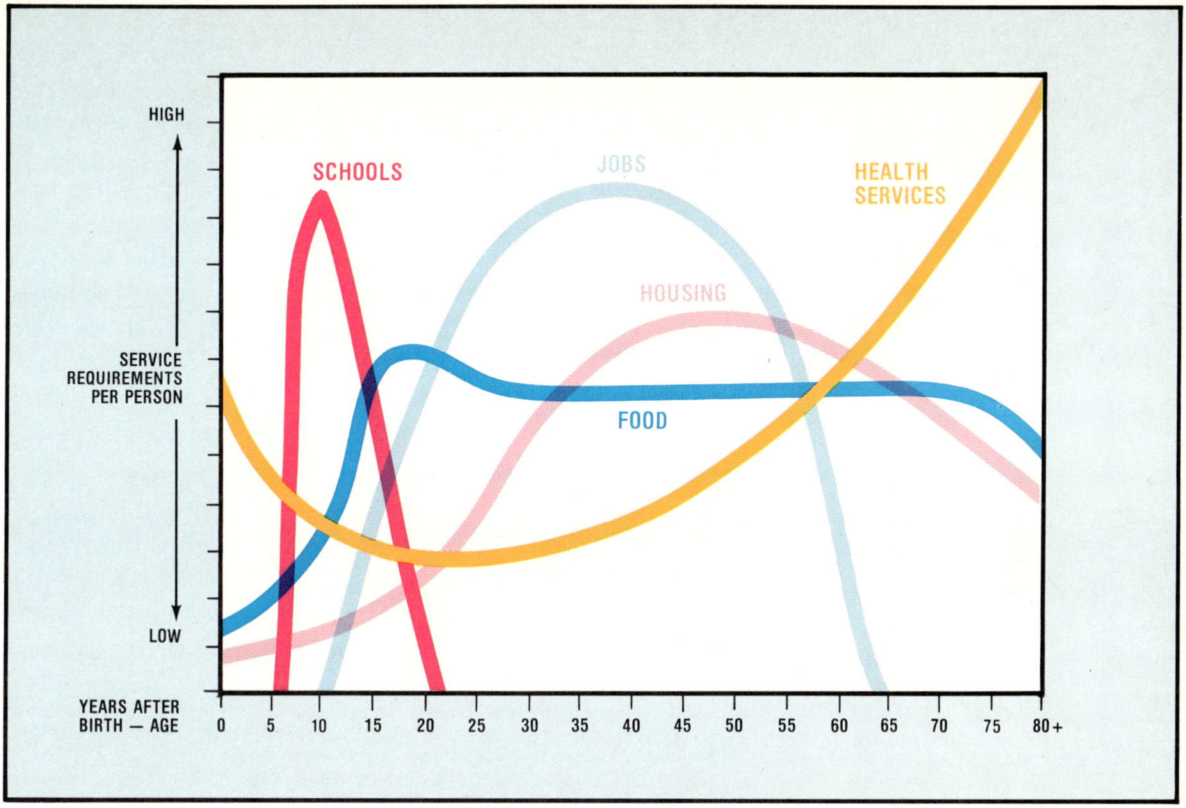


**The service requirements** that occur during a typical human life span are illustrated in this chart. In addition to health services, the chart shows the need for schools, jobs, housing, and food. (World Population Bulletin)

many centuries, people had to fight animals to continue to live. Eventually, however, humans won the battle, largely by using technologies. Today, in virtually all parts of the world, humans dominate animals.

However, smaller organisms continue to do relatively large amounts of damage. Germs, virus organisms, and parasites all live on human tissue. All these are tiny organisms. A germ is any bacteria or other organism that attacks health. A virus is a super-tiny organism. A parasite is any living thing that draws life from another living thing. Many parasites live by drawing food and strength from human or animal blood. All of these organisms are said to be *microscopic.* Microscopic simply means that a microscope is needed to see a given object. This means that the object is too small to be seen with the naked eye. Science and technology have made progress in combating these threats to health. However, diseases can continue to result from infections by microscopic organisms.

Humans have made headway in their battle against microscopic enemies, or *microorganisms.* An earlier chapter describes the black death, or the bubonic plagues, that swept through Europe during the Middle Ages. Individual epidemics killed as much as 50 percent of the population. This has not happened in the twentieth century. Epidemics are monitored closely, and scientists work to find cures or controls. Control over disease is one of the reasons given for the rapid increase in the world's population. This is also one of the reasons why infant mortality rates are higher in countries that have less health care.

Despite continuing progress in fighting diseases, the battle never seems to be completely won. Up to this time, for example, there is no cure for the common cold. One possible reason is that the germs that cause colds keep changing their own physical makeups. That is, as medicines or other treatments develop, the germs acquire *immunity* to them. Immunity means that organisms are not affected by a condition or treatment.

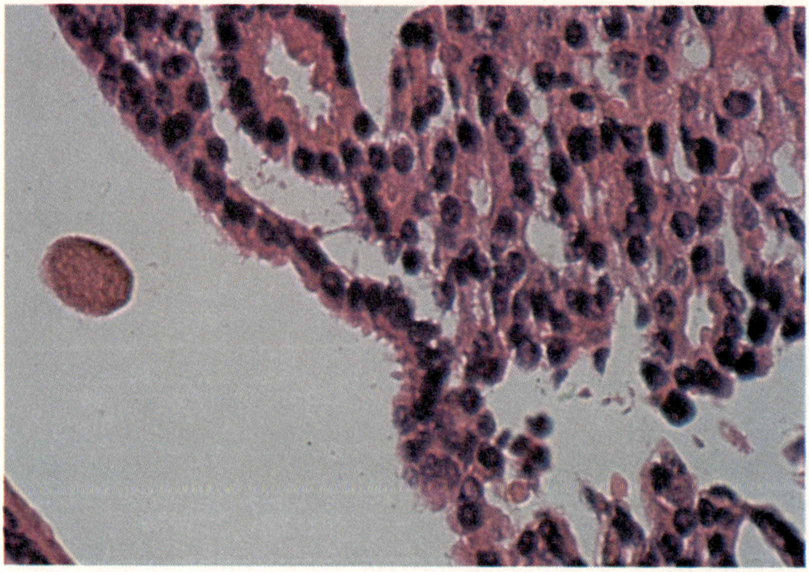

**Microorganisms** are often the greatest enemies of human health. (© Harry J. Przekop, Jr./Nawrocki Stock Photo)

Some medicines can make people immune to certain diseases. Germs can also become immune to medicines.

Another factor is that, as some diseases are conquered, others become greater threats. Also, as people live longer, they become subject to different types of diseases. As one example, the number of deaths from cancer has been increasing in comparatively recent years. Cancer is a disease that destroys human tissue. When tissue destruction reaches a critical level, the body dies. Scientists studying this disease point out that people have been getting older. Most people who contract and die from cancer are in the older segments of the population.

## Trauma

In general terms, *trauma* refers to any kind of pain or suffering. In health care terms, trauma refers to injuries to the body. These injuries most frequently result from accidents. Other causes of trauma include homicide and suicide. Homicide means murder. That is, one person kills, or attempts to kill, another person. In suicide, a person attempts to take his or her own life. If homicides and suicides are successful, people die. Often, injuries result from attempted homicides and suicides. In any case, all three causes—accidents, homicides, and suicides—come under the general heading of trauma.

**Trauma** is common after the body has been through an accident. The nation's highways are particular sources of accidents and trauma. (Courtesy New Jersey Division of Travel and Tourism)

In the United States, more than 160,000 people die from trauma each year. Millions more are injured. The greatest single cause of trauma is the automobile. Some 50,000 people are killed on the highways each year. Almost 1.7 million others are injured.

Care for trauma victims has been a major concentration for health care technology in recent years. In almost every city, one or more hospitals have been designated as trauma centers. These are special facilities staffed by people trained to take care of victims of accidents or violence. All designated trauma centers are staffed around the clock. Many are served by helicopters that rush victims to the centers from the scenes of accidents.

Trauma care technology is based on scientific research about survival from injuries. Research shows that survival rates for victims increase as the time between the accidents and care decreases. That is, the sooner victims can be brought to a trauma center, the greater their chances will be to survive. Survival chances are best if trauma care is applied within one hour of injury.

## Environmental Problems

Technologies are developed in related groups. This is basic to the nature of technology. That is, scientific or technical knowledge, once uncovered, can be used in many ways. Thus, systems of technology have evolved. These systems consist of multiple, related technologies that can affect one another. Sometimes these relationships among technologies deliver benefits. Other times, there can be problems.

Some of the problems of rapid technology development have come in the form of environmental problems. When a new technology is developed, its benefits are seen immediately. Sometimes, problems do not surface until years later. Some examples:

- A new medicine may cure a disease. People hail it as a major benefit. Years later, side effects are discovered that are harmful to human health.

- New methods are developed for processing food. After tons of the new food have been consumed, an unforseen health hazard may be uncovered.

- The automobile, in its early days, was believed to be cleaning up the environment. In New York City alone, many thousands of horses were used for transportation. In 1908, the sanitation department in New York City removed some 12 tons of manure per day from the streets. The automobile was seen as putting an end to pollution. More than 50 years

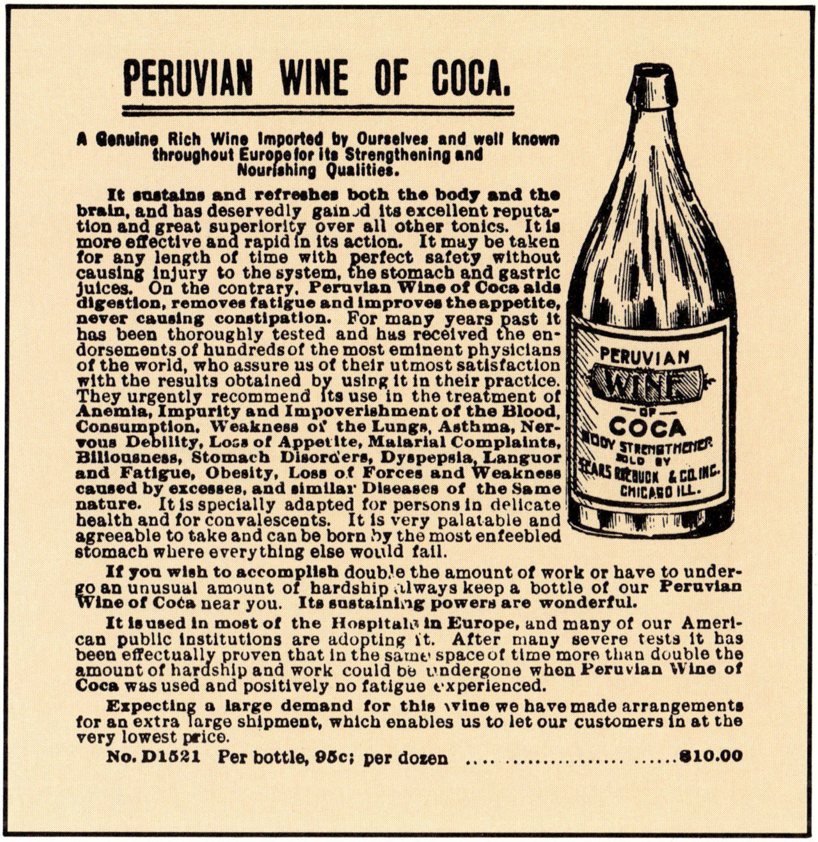

PERUVIAN WINE OF COCA.

A Genuine Rich Wine Imported by Ourselves and well known throughout Europe for its Strengthening and Nourishing Qualities.

It sustains and refreshes both the body and the brain, and has deservedly gained its excellent reputation and great superiority over all other tonics. It is more effective and rapid in its action. It may be taken for any length of time with perfect safety without causing injury to the system, the stomach and gastric juices. On the contrary, Peruvian Wine of Coca aids digestion, removes fatigue and improves the appetite, never causing constipation. For many years past it has been thoroughly tested and has received the endorsements of hundreds of the most eminent physicians of the world, who assure us of their utmost satisfaction with the results obtained by using it in their practice. They urgently recommend its use in the treatment of Anemia, Impurity and Impoverishment of the Blood, Consumption, Weakness of the Lungs, Asthma, Nervous Debility, Loss of Appetite, Malarial Complaints, Billousness, Stomach Disorders, Dyspepsia, Languor and Fatigue, Obesity, Loss of Forces and Weakness caused by excesses, and similar Diseases of the Same nature. It is specially adapted for persons in delicate health and for convalescents. It is very palatable and agreeable to take and can be born by the most enfeebled stomach where everything else would fail.

If you wish to accomplish double the amount of work or have to undergo an unusual amount of hardship always keep a bottle of our Peruvian Wine of Coca near you. Its sustaining powers are wonderful.

It is used in most of the Hospitals in Europe, and many of our American public institutions are adopting it. After many severe tests it has been effectually proven that in the same space of time more than double the amount of hardship and work could be undergone when Peruvian Wine of Coca was used and positively no fatigue experienced.

Expecting a large demand for this wine we have made arrangements for an extra large shipment, which enables us to let our customers in at the very lowest price.

No. D1521 Per bottle, 95c; per dozen .... ................. ......$10.00



**Substances that abuse the body** are readily available and have been for some time. To illustrate, this tonic was advertised in a catalog published in 1897.

later, it was discovered that the gases emitted by automotive engines were polluting the air.

- The automobile has been heralded as the ideal method of personal transportation. It is said that civilization in the United States has been built around the automobile. Yet, the automobile kills some 50,000 people annually and injures some 1.7 million others.
- When it was introduced, asbestos was hailed for its benefits. Asbestos is fireproof. Use of this material greatly improved the safety of homes and other buildings. Asbestos was used in cars and ships as well. Eventually, it was discovered that asbestos greatly increased the rate of lung cancer among persons working with this material.
- Nuclear power plants provide more than 10 percent of all the electric power used in the United States. In the Midwest, more than 20 percent of the electricity comes from nuclear

plants. The nuclear industry has compiled an outstanding record of safety, producing fewer accidents than any other energy-related business. However, major problems have developed in methods for disposing of the waste materials created by nuclear reactors. The fuel and equipment to be removed from nuclear reactors retains dangerous levels of radiation. Radioactivity of some materials will continue for more than 100,000 years.

The point is that all people are surrounded by health threats, all the time. Health can be endangered by air, food, water, or even the content of the earth. Governments are passing laws and taking actions to protect the public from environmental threats. However, these problems will probably never be eliminated completely.

## Psychological Problems

People need one another. That's the main point of urbanization and civilization. Very few individuals can meet all of their own needs. Rather, people move into communities where they can help supply each other's requirements.

But people also create problems for each other. Modern society creates pressures on people who live in it. People who

**People need one another.** However, people together can also create tensions for one another.

live in big cities are affected particularly. These people argue with each other, cause mutual tensions, and engage in violence. Many individuals can't take these pressures. *Psychological illness* has become a major threat to human health. Psychological illnesses are related to mental conditions—conditions of the mind.

Psychology is the study of the human mind. Psychological problems occur among people who cannot deal with, or cope with, the conditions in their lives. People with psychological problems need help. An important part of the medical profession—and of health care technology—lies in psychological care.

However, many people seek help that only tends to worsen their problems. Psychological pressures are major causes of drug abuse and alcohol abuse. Together, uses of drug and alcohol are called *substance abuse.* Substance abuses add up to a major threat to the health of modern human beings.

## Genetic Conditions

Nature also causes some health problems for people. Some 5 percent of all newborn babies have some major mental or physical handicap. In addition, between 10 and 15 percent more have handicaps that are considered minor.

Most of these handicaps are passed on from parents to children. These types of handicaps are known as *genetic defects.* Genetic refers to any traits passed on from parents to children. The term genetic refers to genes. Genes are the units within the body that pass on traits of parents through reproductive processes.

Millions of handicapped people require special health care facilities and methods. Enabling handicapped persons to live normal lives is among the vital targets of health care technologies.

In addition, health care technologies are beginning to analyze the genetic traits of parents. Specialists can now warn prospective parents about traits that will cause problems in children. In many cases, corrective measures can now be taken. Children who would otherwise be severely handicapped can develop normally if these corrective measures are applied.

**Pharmaceutical technology** continuously seeks methods for curing diseases and improving health. (Courtesy National Heart, Lung, and Blood Institute)

## TECHNOLOGIES OF HEALTH CARE

*Pharmaceutical* technology deals with the chemical makeup of the body and with the development of chemicals that affect the functions of the body. Pharmaceuticals are medicines that help the body to overcome or prevent illness. Pharmaceuticals represent a major area of health care.

Pharmaceuticals tend to be developed in a cycle. This cycle corresponds with identification of a disease and development of a cure or a control. During the 1950s, for example, tens of thousands of people, particularly children, were stricken with polio each year. Polio is another name for poliomyelitis. This is a disease that is caused by a virus, a tiny organism that can be seen only with powerful microscopes. Persons who contract poliomyelitis can be crippled severely. Many persons died from this disease.

When a disease injures or kills people, the pharmaceutical industry mobilizes to find a cure. The case of poliomyelitis was typical. For many years, government agencies, private charities, and pharmaceutical companies invested many millions of dollars in research. Ultimately, the cause of polio was identified—the disease is caused by viruses. Further research followed to find medicines that would destroy the virus.

Eventually, two separate groups of scientists came up with *vaccines* to *inoculate* against poliomyelitis. A vaccine is a chemical (medicine) that helps the body build resistance to diseases. Inoculation is the technique for giving a vaccine to people. People who have been inoculated are said to gain *immunity.* Immunity means that people have gained the power to resist a disease.

The cycle followed in developing pharmaceuticals, then, begins with recognition of a need. Following recognition, research is aimed at finding the cause of a disease. When the cause has been discovered, research is aimed at finding a cure (or prevention). Once an apparent cure has been found, further research and testing take place. The testing is aimed at making sure that the pharmaceutical is safe for people. Once safety is established, further research is still needed. This research is aimed at finding ways of producing enough of the pharmaceuticals to meet needs. In addition, the production research involves techniques for assuring the quality of the medicines that are produced.

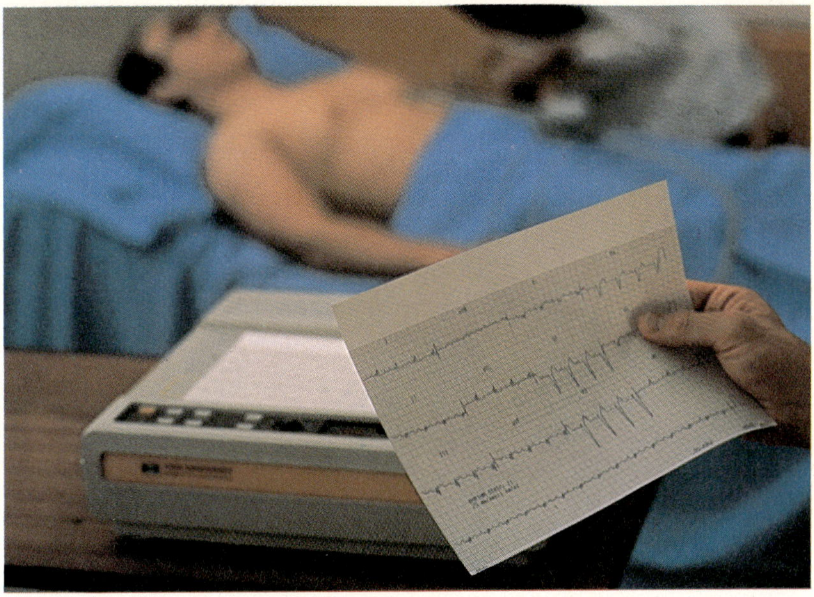

**Diagnostic technology** includes electronic machines that monitor and report upon the body's vital signs. (Photo courtesy of Hewlett-Packard Company)

## DIAGNOSTIC TECHNOLOGY

*Diagnosis* is the process of figuring out why something happens—the cause. *Diagnostic technology* takes in the methods and equipment used in finding the causes of human illness. Diagnosis is an advanced—and also rapidly growing—health care technology. Diagnostic techniques can be grouped in three general categories:

- Vital Signs
- Chemical
- Physical.

### Vital Signs

The vital signs have already been identified. They are pulse, respiration rate, blood pressure, and temperature. Diagnostic technology has developed a wide range of instruments, including electronic devices, for this purpose. These devices both measure and monitor life signs. The electronic units can be set up to sound alarms when a person's life signs show danger signals.

Automatic diagnosis and monitoring systems are used for many types of illnesses. For example, people who have had heart attacks are monitored continuously for signs of further problems. Also, the vital signs of persons who have been through surgery are normally monitored during the recovery period.

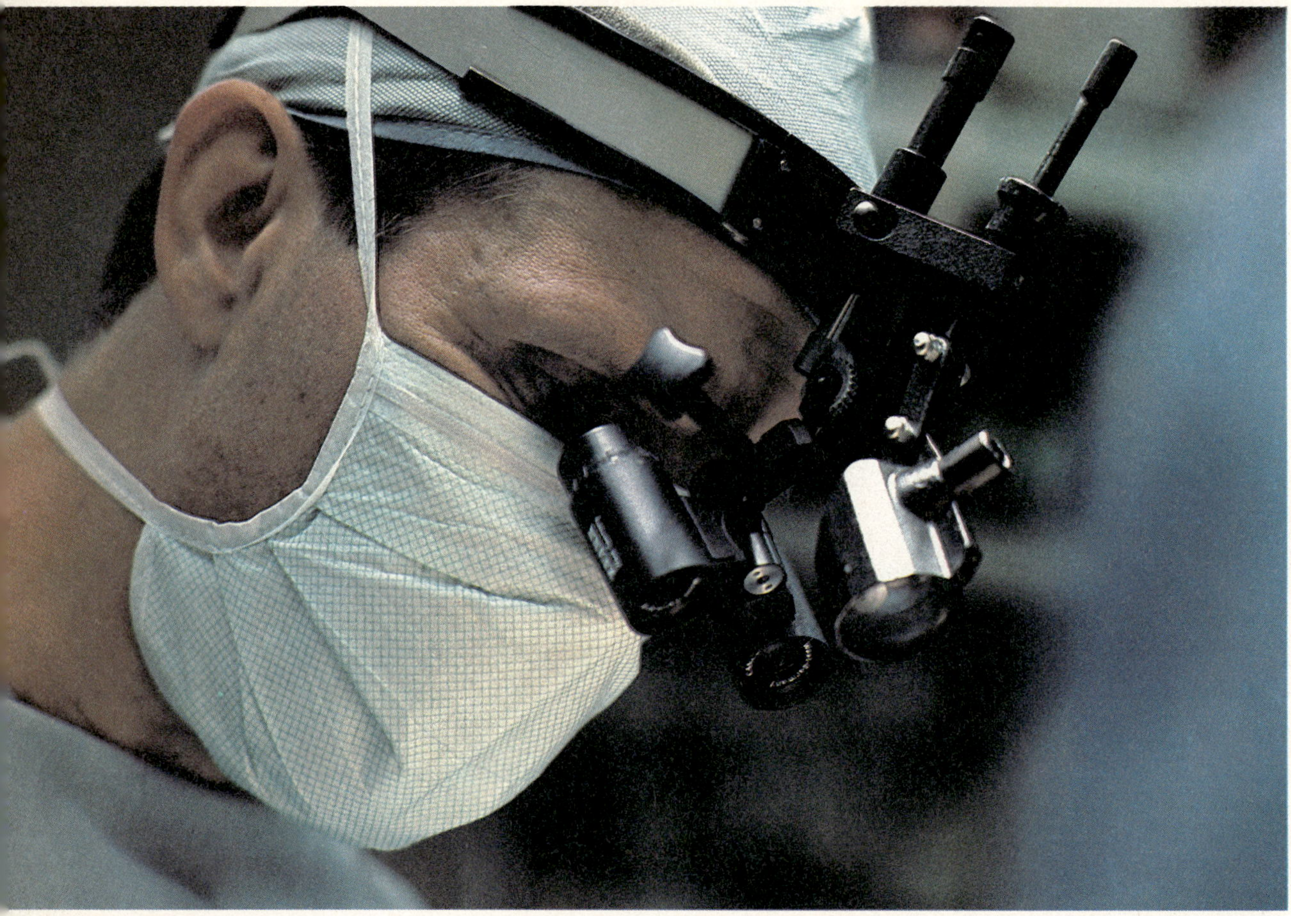

**Delicate surgery** is performed with special optical and lighting instruments. (Charter Medical Corporation)

Surgery is any type of medical procedure that involves cutting into the human body. Recovery is the return of the body to normal health.

## Chemical

Technologies are available, some using computers, to analyze the chemistry of the human body. Chemical diagnosis can be extremely detailed and can reveal many problems even before vital signs are affected.

Many of the chemical analyses are performed on the outputs of the body's excretory systems. Tests can be applied to the breath, to liquid wastes, and to solid wastes. The amount and chemical makeup of body wastes become major indicators of the state of health.

Another major area of chemical diagnosis is the blood. An extremely small amount of blood can be used to perform literally dozens of tests. These tests analyze the functioning of the body from which the blood came. Blood chemistry tests indicate the state, or operating condition, of most major organs of the body.

## Physical

Physical examinations of bodies also provide major diagnostic opportunities. One of the oldest, and still one of the most valuable, physical diagnosis tools is the *x-ray.* X-rays are a form of energy. X-rays have energy levels that make it possible for them to pass through solid objects, including human bodies. X-rays can be recorded on special film. This film shows differences in the types of material through which the x-rays have traveled. Thus, x-ray pictures of human bodies, called *radiographs,* show the bone structures and internal organs.

X-ray pictures are excellent methods for finding broken bones. X-rays also show unusual conditions within the body, such as tumors or other growths.

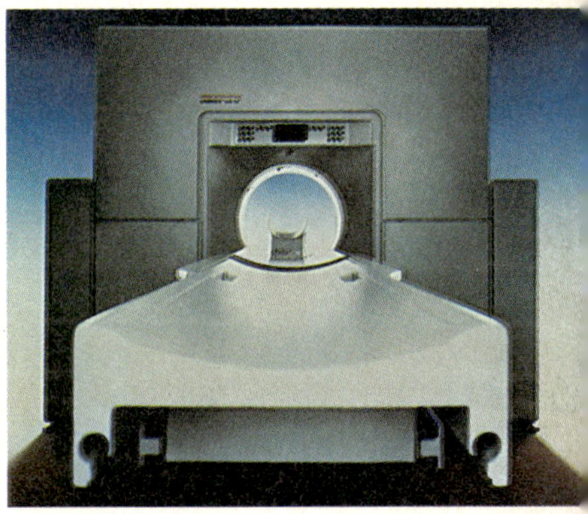

**X-ray examinations** gained new precision and diagnostic value with the introduction of computer controlled examination equipment. (Picker International)

Until the discovery of the x-ray in 1896, surgery was the only method of finding out what was happening within the body. The availability of the x-ray changed almost everything about medicine. For the first time, physical diagnosis could be performed without *intrusive* procedures. Intrusive means to intrude, or to break into something. Intrusive diagnosis involves cutting into the body—surgery.

With the advent of x-rays, techniques for non-intrusive diagnosis were also improved. That is, an x-ray would show that something was wrong. Surgeons could then perform *exploratory* operations to diagnose patients. An exploratory operation is one in which doctors use surgery or special instruments to examine the insides of the body. In addition to surgery, intrusive diagnosis can include the use of needles to remove samples of tissue from internal organs.

As with other forms of technology, x-ray techniques develop their own side effects. The energy generated by x-rays can destroy human tissue. In the process, the body can contract a form of deadly disease, cancer. X-ray exposure and cases of cancer have been linked. Upon discovering this relationship, many uses of x-rays were discontinued or limited. X-ray examinations are still a major form of diagnosis. However, this method is now used with great care.

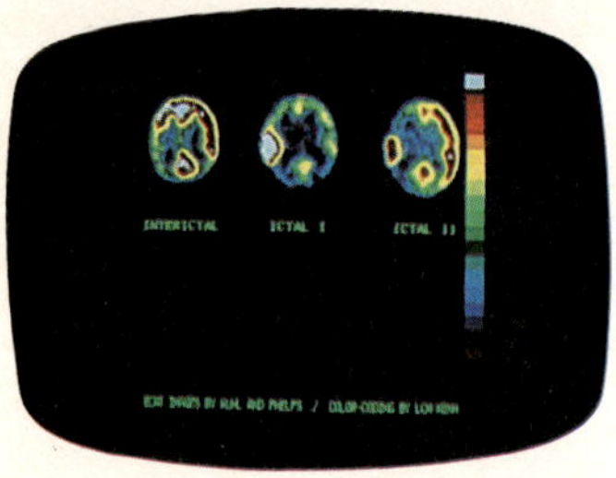

**Computer-developed images** of x-ray examinations help pinpoint conditions and possible problems within the body. (Courtesy National Institute of Health)

Among the controls applied to x-ray examinations has been the use of computers to limit and direct areas of examination. Computer-control techniques now make it possible to take three-dimensional diagnostic pictures of conditions inside the body. A three-dimensional picture shows the full depth, or roundness, of an object. Normal pictures are two-dimensional. The dimensions shown are the height and width. Adding depth makes it possible to diagnose problems more accurately.

Other methods, using special forms of light or sound, are also making instant diagnosis possible. These methods produce the same kinds of pictures as x-rays. However, the types of radiation used are not believed to damage the body.

## TREATMENT

Treatment is the actual care of people. In other words, treatment is the main form of health care for persons with illnesses. The other important form of health care is health maintenance, or illness prevention. Health maintenance is discussed in the chapter that follows. Treatment includes:

- The methods used by physicians in treating illness
- Equipment for caring for and curing illness
- Facilities, including medical offices, hospitals, and rest homes.

The methods used in health care represent the research and experience of physicians. Vast bodies of knowledge have been accumulated. Doctors must read and attend special lectures on new methods continuously.

Much of the equipment used in the treatment of illness is the same as that used in diagnosis. Included are x-ray machines and the vital signs monitoring equipment. Also included are special machines that duplicate the functions of the human body. Doctors now have access to artificial kidneys, artificial lungs, and even artificial hearts.

Facilities for care include hospitals and other, specialized units. Recently, for example, specialized facilities have been set up to treat patients without requiring hospitalization. Even surgery of some types is done on patients who then return home for health care. The methods and equipment used for health care are important areas of technology. These technologies are helping people to keep their health and to regain health when they become ill. Detailed study of hospitals and hospital equipment is not handled in the book. However, the chapter that follows does contain more information about health care and health maintenance methods.

## Summary

Technology is the result of human effort. People think, plan, build, and use technology. Therefore, it makes sense that the area of human health is an important technology. Health care is a system of technologies aimed at keeping people healthy and caring for sick persons.

The human body can be compared to a system of technologies. The circulatory system distributes blood throughout the body. The respiratory system delivers oxygen to circulating blood. The nervous system allows the body to react to external conditions. The muscular-skeletal system gives the body its structure and ability to move. The digestive system changes food to energy. The excretory system eliminates body wastes. The reproductive system involves the development of human life. And the endocrine system reacts to and stimulates all of the other systems. Thus, the body is a group of related systems just as health care technology is a system of related technologies.

People can monitor a body's systems by measuring vital signs. A pulse tells how fast a heart is beating. Blood pressure measures the force of blood as it travels through the arteries. The respiratory rate notes the number of times per minute a person breathes in and out. Finally, a variation in the body's temperature can indicate illness.

## Technology Terms

health care
health
life expectancy
life span
circulatory system
vessels
arteries
veins
circulation
respiratory system
respiration
carbon dioxide
nervous system
nerves
stimuli
impulse
muscular-skeletal system
skeleton
contracting
digestive system
excretory system
excretion
nitrogen
reproductive system

Change occurs continuously in human bodies. An infant must depend on other humans to survive. Once the infant survives the first year, his or her chance for survival increases dramatically. For the next several years, the human body is generally healthy. It is not until the late thirties or forties that the body begins to decline. In old age, the body strength decreases, and people need treatment for illness more frequently.

Several things threaten the health of people. Diseases, trauma, environmental problems, psychological problems, and genetic conditions are threats to human life. Health care technology is needed to treat people if they are affected by these problems. Beyond treatment, health care technology is being applied to discover ways to prevent these problems.

Several technologies, vast in themselves, make up the system of health care. Pharmaceutical technology develops chemicals that affect body functions. Once a problem is identified, pharmaceutical technology sets to work to discover a cure or prevention through chemicals. Diagnostic technology seeks causes of illness. Diagnosis can be conducted through a study of a body's vital signs, chemical makeup, or physical condition.

Treatment includes various technologies that serve people when they are ill. The methods a doctor uses in treating illness are a part of health care technology. Equipment for caring for and curing illness represents another part. Even structures play an important role in treatment. Facilities to treat people include medical offices, hospitals, and rest homes.

The system of health care technologies is complex and fast-growing. People want to know as much as possible about the human body. Therefore, knowledge in this area is accumulating at a rapid rate.

## Using Technology

1. Determine your pulse rate after you have exercised and, again, when you are resting. Compare the results. Also, measure your body's temperature several times during a day. Did your temperature stay the same? Jot down your answers to share in class.

2. In the 1950s, government health programs were set up to immunize school-age children against several dangerous diseases. In the 1970s, most of these programs were discontinued, and parents became responsible for immunizing their children. Today, however, studies show that many children have never been immunized against diseases such as polio and whooping cough. What might be the result if this situation continues to exist?

3. It has been said that a person who smokes one pack of cigarettes per day cuts off an average of seven years from his or her life. What, then, is the effect upon the life expectancy of a person living in the United States who smokes two packs of cigarettes per day?

endocrine system
endocrine
hormone
adrenaline
life sign
vital sign
pulse
artery
blood pressure
respiratory rate
stethoscope
thermometer
infant mortality rate
treatment
microscopic
microorganism
immunity
trauma
psychological illness
substance abuse
genetic defect
pharmaceutical
vaccine
inoculate
immunity
diagnosis
diagnostic technology
x-ray
radiograph
intrusive
exploratory

# 11 HEALTH MAINTENANCE

## Your Learning Job

After you have read this chapter and completed the work assignments, you should be able to:

- ☐ Explain the difference between health care and health maintenance.
- ☐ Describe the difference between a health maintenance organization and traditional medical insurance.
- ☐ Explain the purpose of a physical checkup.
- ☐ Name several common examples of preventive medicine.
- ☐ Define holistic medicine and describe techniques used in this approach to health maintenance.
- ☐ Name the elements of a proper diet.
- ☐ Outline an exercise program that would promote good health.
- ☐ Explain how physical therapy programs help people.

## TECHNOLOGIES FOR HEALTH MAINTENANCE

**Health maintenance** programs are designed to enable people to protect and enjoy the natural health of their bodies.

Health care involves treatment and attempted cures for illness. Health maintenance, on the other hand, deals with people who are enjoying normal, sound health. The idea is to keep the health you have, treating yourself while you are well instead of waiting to get sick. Health maintenance has been likened to preventive maintenance in automobiles. If you own a car, you take it into the shop from time to time to perform certain maintenance. For a car, maintenance includes lubrication, changing of oil, and checking of the working parts.

These same principles apply in maintaining the health of people. Diagnostic techniques are used. These include x-rays and chemical analyses. The idea is to monitor the condition of the body. Measures can be taken to build and maintain sound body functions. Some of the techniques and technologies used in health maintenance include:

- Health maintenance organizations
- Checkups
- Preventive medicine
- Holistic medicine.

## Health Maintenance Organizations

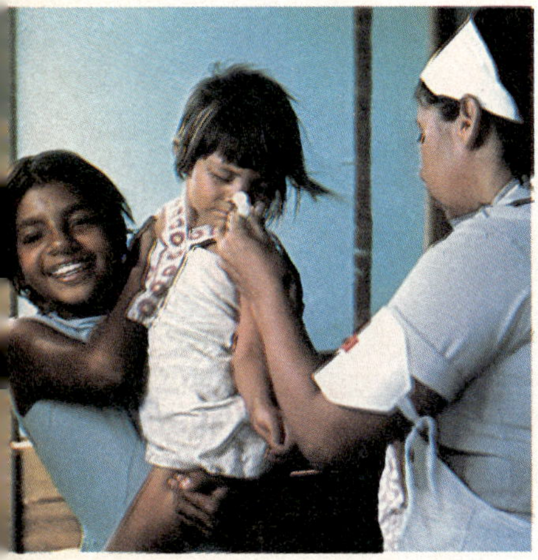

**Regular medical checkups** are an important part of health maintenance. (American Red Cross)

A *health maintenance organization* provides a type of medical, or health, insurance. Traditional health insurance pays medical bills for the treatment of sick persons. Persons who feel ill go to doctors. The doctors submit their bills to the insurance companies. Insurance usually covers all, or most, medical costs. Recently, medical insurance has been expanded to include annual (yearly) physical examinations. A physical examination is usually considered a part of health maintenance. However, health maintenance organizations go further.

A health maintenance organization provides total care for members. Persons who belong to health maintenance organizations are considered members, not patients. Doctors and technicians who work for health maintenance organizations take care of members continuously. That is, they don't wait until people are sick to provide medical attention. Instead, regular appointments are made for checkups.

Members are encouraged to take part in programs involving diet, exercise, and self-diagnosis. The idea is to have the members monitor and protect their health. This should minimize illness. In many cases, it has been proven that members of health maintenance organizations spend fewer days, on the average, in hospitals than other citizens. The point is, again, to treat people while they are well rather than waiting for them to get sick.

## Checkups

The most common form of health maintenance checkup is the annual physical examination. Under some health maintenance plans, people are required to go through an extensive examination every year. The annual checkup includes a complete physical examination. The physician checks for the member's vital signs and interviews the patient to identify any physical changes. Further, a checkup usually involves an extensive series of chemical diagnostic tests. Chest x-rays to screen for diseases of the heart and lungs are usually included. Under some plans, x-rays are taken only every other year, to minimize exposure.

The steps in a physical examination are usually geared to help physicians to find anything wrong with a patient. The idea is to find potential problems before illness occurs. Then, corrective measures can be taken before illnesses get worse.

In particular, annual checkups look for physical changes in patients. Patients who gain or lose weight are asked about reasons for these changes. Overweight can be dangerous. However, loss of weight can also indicate changes of body conditions that should be monitored. The same is true of changes in readings of vital signs. Rising blood pressure is usually monitored closely.

**Preventive medicine** programs include regular physical checkups. This photo shows the use of special equipment for hearing tests. (© Larry Brooks/Nawrocki Stock Photo)

One of the advantages of an annual checkup lies in the continuing history provided for a physician. If a patient is examined each year, the physician can check for changes that can indicate potential problems. If any such changes occur, the doctor can put the patient onto a program of monitoring or *therapy*. Therapy is another term for treatment. These monitoring and therapy programs are often used for patients with conditions that are not yet serious. That is, changes have been noted. However, patients are not yet sick enough to require medical treatment, surgery, or hospitalization. Instead, the idea is to deal with the minor changes before they become major.

## Preventive Medicine

Health maintenance measures are part of the technology for *preventive medicine*. In addition, preventive medicine also includes medication to avoid diseases.

Typical preventive measures include the use of vaccines to avoid diseases. For example, all children are usually required to be vaccinated against certain childhood diseases before they start school. Because of programs of this type, many diseases that used to kill millions of children are now practically dead themselves. This is true, at least, in the United States. In other parts of the world, millions of children still die from such childhood diseases as diphtheria, whooping cough, and polio.

Inoculation programs are also used for many adults. For example, it has become common in recent years to inoculate older people and persons with respiratory diseases against the flu.

### Holistic Medicine

*Holistic* means whole. *Holistic medicine,* then, refers to care of the whole person. Holistic medicine uses a combination of techniques to provide for the mind, body, and spirit. Holistic medicine encourages people to think their way to sound health. Minimum use is made of drugs, surgery, or other medical-type treatments.

One of the areas of research being followed in holistic medicine is known as *biofeedback*. Biofeedback combines awareness of physical conditions with positive mental concentration. By being aware of conditions, it is felt that individuals can do something about them. People are trained, sometimes with electronic aids, to detect their body's physical conditions. The idea is to trigger natural physical reactions that help the body to correct conditions. This technique and others are used to emphasize concentration on health, rather than sickness.

## PERSONAL HEALTH MAINTENANCE

A major principle of health maintenance is that each individual should have both pride and responsibility where health is concerned. That is, a person should be responsible for following programs that help to maintain good health. As part of the same program, a person should build a neat, attractive appearance. People should be proud of their appearance. This combination of responsibility and pride can help to insure lifelong good health. Techniques used in personal health maintenance include:

- Diet
- Exercise
- Physical therapy.

### Diet

*Diet* is a term that describes everything you eat and drink. Your total food and liquid intake, in turn, affects your health and appearance. It is said often: "You are what you eat."

Emphasis on health maintenance has focused attention on diet. At one time, people used to eat what they enjoyed eating

**Eating the right foods** can contribute to the development and health of the body. A balanced diet should include selections of foods for their nutritional values.
(Texas Department of Highways/Travel and Information Division)

or whatever was available. Considerable diet-related illness occurred because people, in years past, could not get vegetables during the winter. People's diets, at one time, included food that led to the gaining of weight. People looked healthy and chubby. But their health suffered because their bodies did not get the right fuel and *nourishment*. Nourishment is a term that takes in the chemical building of the body. Muscles need one kind of food, bones need another, and so on. Diets must provide both fuel and nourishment.

**The more active a body is,** the more calories it will burn. Diets should be adjusted to activity levels. (© Paul Degruccio/Uniphoto)

The fuel value of food is measured in *calories.* A calorie measures the amount of heat that can be generated from burning a given amount of fuel. Some foods produce more calories than others.

Your body needs only so many calories of input to sustain itself. The number of calories you need depends upon how active you are. For example, a professional athlete may burn 4,000 or more calories a day. On the other hand, an average person of between 30 and 40 years of age may burn only 1,000 calories a day. Teenagers, typically, burn anywhere from 1,500 to 3,000 calories per day.

The value of foods in your diet depends upon the calories they contain and also upon their nutritional properties. Nutritional values of foods are determined by the ingredients they contain. Food ingredients, in turn, can be classified according to their chemical makeup. Chemical ingredients of foods fall into three distinct categories:

- Carbohydrates
- Fats
- Proteins.

Your body needs a balanced intake of these three chemical ingredients. You may hear about so-called new diets that eliminate one of these ingredients while stressing others. It is true that you do have to control the amount of carbohydrates, proteins, and fats you eat. However, you do need all three of these elements in your diet. In addition to these, you also need some other ingredients, such as water, vitamins, minerals, and fiber. These other elements of your diet are discussed below.

**Carbohydrates.** *Carbohydrates* provide high levels of fuel, but very little nutrition. Carbohydrates are commonly known as sugars and starches. Foods high in sugar content include candies and many soft drinks. Food high in starches include potatoes and bread.

As indicated above, your body needs some carbohydrates. However, your diet quickly can reach a point where you are getting too much of a good thing. For example, a dessert at the end of the meal might have almost as many calories as everything else that you have eaten. Remember, your body burns only so many calories. If you take in more calories than you need, your body turns carbohydrates into fats. Fats add weight and can impair your health.

**Fats.** *Fats* are similar in chemical makeup to carbohydrates. The main difference, as far as diet is concerned, lies in calorie content. Fats contain twice as many calories, by weight, as carbohydrates.

Fats include the white, slippery substance that you see at the edges of certain meats, such as steaks. Butter and margarine are also almost pure fat, as are many cooking oils. Thus, fried foods, such as french fried potatoes, represent a combination of carbohydrates and absorbed fats.

Some amount of fat should be included in every balanced diet. For some people, fatty foods are considered healthy. For example, Eskimos eat comparatively large amounts of fat. They need this because their bodies burn large numbers of calories to keep warm in cold climates. Active athletes also burn large numbers of calories and may not be affected by fats as much as

**Sports and competition** require physical fitness and careful attention to diet. Many athletes consume extra carbohydrates just prior to competition. (Hawaii Visitors Bureau)

other people. However, the average person should limit fat intake in his or her diet.

**Proteins.** *Proteins* provide the same calorie content as carbohydrates, according to weight of food consumed. However, protein is also rich in nutrients that build body tissue. In particular, protein intake is easily converted to muscle tissue. Some of the protein you eat actually is muscle. That is, if you eat meat substances, you are consuming animal muscle. This food then converts directly to muscle in your own body. A number of plant foods are also high in protein. Included are most legumes, such as beans and peas.

**Other diet elements.** Your diet should also provide supplies of water, vitamins, minerals, and fiber. Water is a necessity of life. You can actually die in a shorter time if you deprive your body of water than if you stop eating. Your body is approximately two-thirds water. Therefore, your diet should keep replenishing its liquid content. Further, if you drink more water than you need, your body will dispose of what it doesn't absorb.

**A balanced diet** can travel with you to points of recreation or on long journeys. The food you eat should be a continuing concern. (Florida Division of Tourism)

*Vitamins* are chemicals that cause other chemicals to be used. Your body needs certain amounts of vitamins to help use the food you eat. There are vitamins in most foods you eat. If you maintain a balanced diet, you will probably get all of the vitamins you need.

However, many people take vitamin *supplements* as additions to their diets. Most food and drug stores sell vitamin supplements. Your body may be able to benefit from some of these substances. However, it is not a good idea to take vitamins without discussing your diet with a doctor or nutrition expert. In general, vitamins are good. But some vitamins, if you take overdoses, can be harmful.

*Minerals* are chemical elements found in nature. Examples include iron, copper, and calcium. Your body needs some supplies of minerals. For example, the iron in your blood helps to absorb oxygen. The calcium in the milk you drink helps build bone structures.

As with vitamins, mineral supplements are also available in food and drug stores. Similarly, you should know what you are

**Some foods,** such as milk, contain many diet elements required by a developing body.

taking—and why you are taking it. Consult a doctor or nutrition expert before you start taking any substance into your body.

*Fiber* is also known as *roughage.* Fibers, in effect, are the foods that push other elements of your diet through the digestive system. Thus, roughage helps the body to eliminate waste properly. Examples of foods that contain fibers include cereal, lettuce, and most fruits.

The idea of diet is to give your body the chemicals it needs. At the same time, your diet should avoid giving your body things that can be harmful. Items that can hurt your body, particularly during growing years, include alcohol, tobacco, and drugs. Unnecessary use of any of these materials is substance abuse. This topic is discussed in the previous chapter. All that needs to be said here is that it doesn't make sense to destroy your body purposefully.

## EXERCISE

Your body is an organism, or mechanism, that is designed to do work. Think back about the definition of doing work. Work is a combination of force and movement. To remain healthy, any human body must apply force and achieve movement. This occurs when the body exercises. Exercise is necessary for every person. Exercise should also be done regularly. A typical recommendation is that every person should exercise strenuously for at least 20 minutes at a time. These 20-minute sessions should take place at least three times a week. Strenuously means that you put a physical strain on your muscles.

**Aerobic exercise** involves movement of the body. This results in an increase of the circulatory and respiratory functions. Sports such as tennis provide aerobic benefits. (Florida Division of Tourism)

When you exercise, you also strain, or put to heavy use, your circulatory and respiratory systems. The muscles that are doing the work require oxygen and fuel from the blood. This means that your respiratory and circulatory systems work harder when you are exercising than when you are resting. So, exercising your arms and legs also exercises your heart and lungs.

In general, there are two approaches to exercising. These are:

- Aerobic
- Isometric.

For either type of exercise program, be aware that preparations should be made. Avoid exercising right after you have eaten. Wait at least 30 minutes after meals before exercising. The idea of exercise is to put a planned strain on your circulation and respiration. Also, you want to control the flow of blood and its oxygen content to your muscles. Immediately after you eat, much of your blood is directed toward your digestion process. If you exercise on a full stomach, your brain and muscles don't receive their full blood supply. After 30 minutes or a little longer, circulation returns to normal.

Before going ahead with exercises involving strain, you should warm up. This can be done with mild exercises or by stretching. For example, runners or other athletes often jog before a race or game. This gets the blood flowing to the muscles that will be affected by exercise. As an alternative, stretching exercises can be used. Assume positions that stretch the muscles to be used. Hold the stretch positions until a mild strain occurs. This also serves to stimulate circulation. Strenuous exercise without a warmup can lead to strained or damaged muscles.

## Aerobic Exercise

*Aerobic* exercise involves movement of the body. Examples include walking, running, and swimming. Other forms of aerobic exercises are calisthenics and dancing.

To be effective, aerobic exercise must be carried on for some period of time. An aerobic condition exists when your heartbeat, or pulse rate, reaches a certain point. To benefit from aerobic

**Isometric exercise** tones the body's muscles. Calisthenics and weight lifting are examples of isometric exercises. (UNRWA)

exercise, you should maintain this heart rate for 20 minutes or more. That is, an aerobic exercise session should continue for at least 20 minutes after you achieve an aerobic heart rate.

The aerobic pulse rate for any individual is determined by his or her age. Some simple arithmetic is involved. Start with the value of 220. Deduct your age. The difference is your maximum aerobic heart rate. Then take 70 percent of the maximum rate. The pulse rate for any individual should be between 70 and 80 percent of the maximum rate, based on age. You should never maintain 100 percent of your maximum rate for an extended

**Exercise is a lifelong** requirement for people who want to maintain the best health possible at any given age. (Courtesy The Salvation Army)

time. However, certain persons, such as athletes, do approach the 100 percent level for short periods.

To illustrate, a 15-year-old person would deduct this age from 220. The difference is 205. Next, multiply 205 by .70. The result is 143.50. The person's aerobic pulse rate would be between 143 and 163. In an organized exercise plan, this pulse rate should be maintained for at least 20 minutes. At least three exercise sessions of this type should be completed each week.

## Isometric

*Isometric* exercise programs are designed largely for muscle toning. Muscle toning is the building and maintaining of strong muscular tissue. In isometric exercises, no special attention is paid to pulse rate. The emphasis is upon the building of physical strength.

Isometric exercises include weight lifting and certain types of calisthenics. In isometric calisthenics programs, push-ups or exercises that pit muscles against each other are often used. To illustrate, place your hands together in front of your chest and push your arms against each other. In a short time, you will feel a strain. Wait a little longer and you will feel warmth in your arms. This is an example of an isometric exercise and its effects. Other isometric exercises involve pushing against walls or other objects.

A popular form of isometric exercise is weight lifting. The idea is to lift or push relatively heavy weights until the arms or legs become tired. When you become tired, you rest briefly. Then you repeat the exercise.

The idea of isometrics is to strain specific muscles in a planned pattern. One specific set of muscles is put under stress at a time. This causes extra blood to flow to those strained muscles. The additional supplies of oxygen help build those muscles.

**A symbol of strength,** through the ages, has been the lifting of weight. Today, weight lifting is recognized as a potential part of a balanced exercise program.

It is possible to devise exercise programs that combine isometrics and aerobics. One approach is to do a series of isometric exercises. Use fewer weights and reduce the strain. Do more repetitions of the exercise, for longer times. Keep up these isometric patterns for 20 minutes or more. When you do, your body will reach an aerobic state.

Whatever type of exercise a person does, appearance is usually one of the reasons. People exercise to feel well. But they also exercise because they want to look well. In a balanced program, health and good appearance can go together.

## PHYSICAL THERAPY

*Physical therapy,* in general, is a means of bringing the body back to health after an illness. Therapy is often recommended for accident victims or people who have had heart attacks. The idea is that the body has been damaged. Careful, controlled exercise helps to rebuild damaged muscles. Also, if muscles have been destroyed, other muscles can be developed to offset the damage.

Physical therapy often involves special types of exercise. In addition, body movements are sometimes aided by special equipment or trained technicians. Technicians specializing in physical therapy are called physical *therapists.*

In addition to accidents and heart attacks, physical therapy is also used to overcome other types of illness. Examples include strokes or paralysis. A stroke is an attack that causes loss of brain function. Usually, the part of the brain that is destroyed controls part of the nervous system. The nerves involved cause movement of the body. A person who has suffered a stroke loses the ability to move part of his or her body. When this happens, the muscles involved lose their tone or strength. This is called deterioration, or *atrophy* of muscles. Special exercise programs help to minimize the effects of muscle atrophy.

The loss of the use of the muscles through strokes or other causes is also called *paralysis*. Physical therapy is also used for persons who have become paralyzed.

## Technologies for Personal Health Maintenance

Historically, people used to get enough physical exercise to keep their bodies toned. When people had to hunt to get their food, exercise was no problem. The same was true when most people worked on farms. In an urbanized society, however, exercise programs become a matter of technology. Major industries have built up around these technologies. Health clubs and spas have

**Recreation and health** together through participation in sports. Sports bring people outdoors and also provide opportunities for exercise. Many adults stay in shape by playing golf. (Florida Division of Tourism)

become a major industry. These are facilities that specialize in exercise equipment and programs. In addition, many specialized exercise facilities have been opened. Examples include clubs for tennis and racquetball. Golf is a long-standing sport that people use for personal health maintenance.

People don't need special facilities to exercise. However, even people who exercise on their own may need special equipment. For example, people who jog as part of their exercise programs should wear special shoes. Without proper shoes, you can hurt your feet if you run on pavement. Also, many people who exercise regularly wear special clothing. This clothing is designed to retain the body heat. When the body is heated during exercise, calories are burned. Thus, exercise can help in weight control as well as in conditioning the heart and other muscles.

As indicated at the beginning of this chapter, these technologies of health care are vitally important. People live longer because they are cared for when they become ill. It is also possible, however, to use technologies to guard against illness. Health care and health maintenance programs can work together to provide both health and happiness.

## Summary

Health maintenance technologies are related to health care technologies. However, health maintenance goals are directed toward healthy people, while health care goals lean toward treatment and cure of illnesses.

Health maintenance techniques include health maintenance organizations, physical checkups, preventive medicine, and holistic medicine. A health maintenance organization provides medical insurance but differs from a traditional insurance program in its approach toward patients. Total care is provided for members. Physical examinations and programs to educate patients offer preventive measures. A physical checkup helps to spot trouble before a person becomes ill. Preventive medicine includes techniques such as vaccines to prevent diseases. Holistic medicine offers new techniques in treating the whole person.

Health maintenance assumes that people should take responsibility for their health. Therefore, many techniques can be used by an individual. People can learn what makes a proper diet. People can learn to exercise in ways that will promote good health. And, people can follow physical therapy programs after an illness or injury to return to good health.

A proper diet consists of several elements: carbohydrates, fats, proteins, water, minerals, vitamins, and fiber. The idea of diet is to give your body the chemicals it needs. Some chemicals, including alcohol, tobacco, and drugs, are harmful to the body.

Because the human body is designed to work, an exercise program is necessary for lasting health. Regular, strenuous exercise for 20 minutes, three times a week, is a recommended minimum. Aerobic exercises involve body movement to raise the heart rate. Swimming and running are aerobic exercises. Isometric exercises are designed for muscle toning. Weight lifting and some calisthenics are isometric exercises.

Physical therapy programs are often recommended for accident victims or people recovering from heart attacks. A typical program would include special exercises, perhaps using special equipment.

People who learn to take care of their bodies enjoy good health. A combination of good health care and health maintenance should help people to live longer, healthier lives.

## Using Technology

1. Analyze a meal you have eaten recently. Would you say that the meal provided enough nutrition? If so, explain what diet elements were included. If not, explain which elements were missing. Write your answer in a short composition.

2. What is your aerobic pulse rate?

3. Suppose that you ate the following foods in one day. Breakfast: eggs, fried potatoes, bacon, toast, orange juice, and milk. Mid-morning snack: candy bar and cola. Lunch: hamburger, french fries, large cola. Afternoon snack: potato chips and dip. Supper: pork chops, mashed potatoes, gravy, green beans, and iced tea. Further suppose that you were not active that day. Based on the following calorie chart, did you eat too much? Too little? Explain your answer in a short composition.

| | | | |
|---|---|---|---|
| eggs | 180 | toast | 190 |
| fried potatoes | 150 | orange juice | 110 |
| bacon | 85 | milk | 150 |
| candy bar | 160 | cola | 145 |
| hamburger | 365 | french fries | 180 |
| large cola | 260 | | |
| potato chips | 230 | dip | 100 |
| pork chops | 620 | green beans | 30 |
| mashed potatoes | 160 | iced tea | 25 |
| gravy | 250 | | |

## Technology Terms

health maintenance organization
therapy
preventive medicine
holistic
holistic medicine
biofeedback
diet
nourishment
calorie
carbohydrates
fats
proteins
vitamin
supplement
mineral
fiber
roughage
aerobic
isometric
physical therapy
therapist
atrophy
paralysis

# IV

# INFORMATION TECHNOLOGY

Through many centuries, information and knowledge developed as by-products of technology. As people developed technologies to meet needs and wants, they learned from their discoveries and inventions. The information and knowledge that people built has been applied to develop new inventions and new progress. In time, information-handling methods have formed a vital technology of their own. This part introduces information technologies and the tool that has evolved to meet information processing needs. This tool is the computer.

Chapter 12 describes information as a human resource. The basic requirements for developing and handling information are covered. The methods described are usable with or without computers.

Chapter 13 introduces the computer and describes how computers work. Also covered are the contributions of computers to information technologies.

Chapter 14 describes the role of computers in changing the way business is conducted. Computers have affected the lives of all people. These effects of computers are reviewed.

# 12

# INFORMATION SYSTEMS

### Your Learning Job

After you have read this chapter and completed the work assignments, you should be able to:

- ☐ Define data, information, and knowledge.
- ☐ Explain how a computer is different from a calculator.
- ☐ Explain the relationships among data, information, and knowledge.
- ☐ Outline the development of computer technology.
- ☐ Describe an information processing cycle.

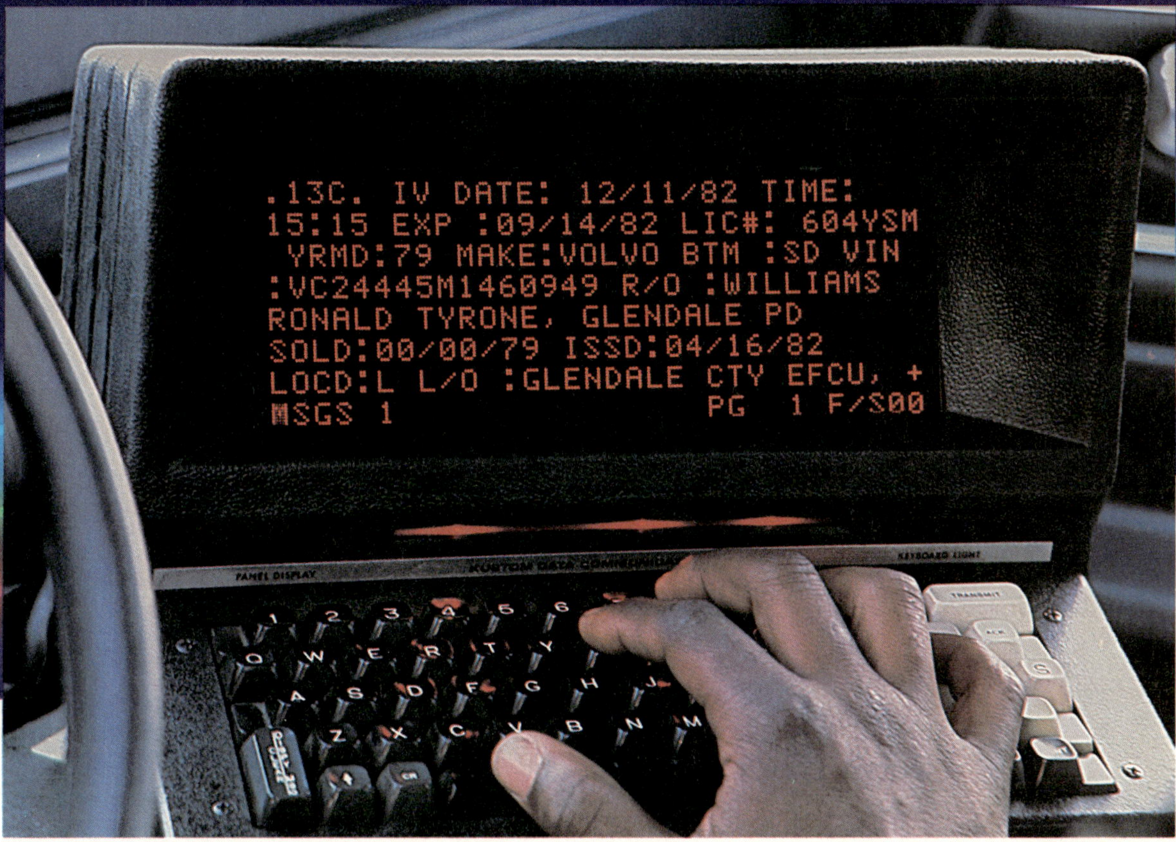


**The need for information** can be urgent. This display, set up in a police car, can contain information of life-or-death importance.

## THE IMPORTANCE OF INFORMATION

The story that follows is not based on a specific company. But the incidents are true. As a matter of fact, this basic situation has been repeated many times.

*The company was growing. Business was excellent. As a matter of fact, growth had been so rapid that a computer became a necessary tool. Management said that, once records were computerized, the company became stronger. Management had more information. The computer could be counted on as a tool to control daily operations and to make decisions about the organization's future.*

*All of that ended one night. Fire wiped out the company's home office. The computer center was destroyed. Even worse, all of the records in the computer center also went up in flames. The loss included the computer programs, as well as the procedures for using the computer. Worst of all, however, all of the information records developed and maintained by the computer were destroyed.*

**Companies have failed** when fire has destroyed computer-maintained records. In today's world, information is a necessity of business life. (Four By Five, Inc.)

*The company had insurance policies covering all of the damage. However, with the computer records destroyed, there was no way to tell exactly how much damage had been done. Eventually, the insurance company made an offer and a settlement. But that didn't matter. Without the computer-produced records, there was no longer a company. The computer had kept all records about customers, and amounts that the customers owed to the company. All government and employee records were also on the computer. Without its computer and computer-produced records, the company simply couldn't continue in business. The organization eventually went out of business.*

Because of experiences of this type, a principle has developed in the computer field: Computer-produced data and information are vital resources for any company. If an emergency develops, the computer-produced data and information **are** the company.

At a different, more general level, it is said that the United States has become an *information society*. Certainly, there is no doubt that information has become a necessity of life for any business or government organization. People who build successful careers depend upon their ability to understand the importance of information and the information technology.

## SOURCES OF INFORMATION

Computers process *data* and produce *information*. Information, in turn, is used by people, who develop *knowledge*. The relationships among data, information, and knowledge are important to an understanding of computers and what they do. Computers accept data and produce information. People use information. It takes the imagination and skill of people to generate knowledge. However, computers can help people in putting knowledge to work.

The key terms in understanding the importance and value of computers, then, include:

- Data are basic, or raw, facts. Examples of data include today's date and the name of your school. There are also many *data items* about you. These include your name and address. Other data about you include your height, weight, age, and color of eyes. A characteristic of data items is that they have little meaning until they are related to one another. As an

**You are the focus** for the collection of data and the development of information. You continue to generate data throughout your life.

example, the data item about the color of your eyes would not mean anything by itself. However, when this data item is connected with your name, it helps to identify you.

- Information, then, is formed by the relating of data items. When your name and the color of your eyes are related, these data items become information. The more data items that can be collected to describe you, the more information there is about you. Information results from the connecting, processing, or relating of data items. *Processing* of data takes place when data items are handled in any way. Examples of processing include the basic arithmetic functions—addition, subtraction, division, and multiplication. In addition, data processing can involve the comparison of two data items.

- Knowledge comes from information. The development of real knowledge requires human brains. That is, people use information and relate collections of information. The connections made by people provide the basis for knowledge. For example, doctors may gather information about how hundreds of different drugs or combinations of drugs act upon a germ. The doctors build knowledge until they come up with an answer—a cure for a disease. Knowledge, then,

**Computations to support scientific experiments** were among the major, early needs that led to development of computers.
(Courtesy Tenneco Inc.)

comes from the use and analysis of information. Examples of knowledge include an understanding of the makeup of atoms and the formation of molecules. This understanding comes from a body of knowledge known as the science of physics. The knowledge results because people put information together. The relationships that result give the information special meaning. The technology of information processing, and the systems of technology that serve people, all result from knowledge.

## COMPUTERS, DATA, AND INFORMATION

The main tool for processing data to develop information is the *computer*. A computer is a collection of connected devices (units of equipment). The assembled computer must be able to handle entire jobs on its own. There are many types of equipment that can process data. For example, a hand-held calculator can add, subtract, divide, or multiply sets of numbers. However, the calculator requires a human operator for each step performed.

The computer is different. A computer can store and follow complete sets of instructions, known as *programs*. Programs are written by people and then entered into computers. Once a computer has a program to follow, the machine can complete entire jobs on its own. Automatically, a computer can accept sets of numbers or words, perform complex processing, and deliver specific results. This entire set of steps is completed without any help or direction from people. Once people prepare the programs, computers can follow these instructions. In some cases, thousands of processing functions can be carried out without any help from people.

## Computer Processing: An Everyday Example

To illustrate how computers process data to produce information, consider what happens in a department store or supermarket. Suppose you and your family buy 40 or 50 items in a supermarket. You push a full basket of food items up to a checkstand. The checker (the person who operates the equipment at the checkstand) handles your purchases one at a time.

**Computer code** is part of the label of most products you buy in supermarkets. Bar codes of the type shown above can be read directly into computers. The data sensed from these special labels are used to record and total your purchases.

Each purchase is passed across the counter of the checkstand. As the food item is handled, a special label on the package is read automatically. This reading takes place as the package passes over a beam of light. The light comes from a special computer-connected device. This reading device is built into the counter of the checkstand. The label on the package consists of a series of lines and blank spaces. These lines and blank spaces form a code—called a *bar code*—that can be read by computers. The bar code on each label identifies the package and its product. This identification is transmitted to a computer in the store.

The computer then operates the cash register automatically. The register prints a description of the product and its price. With this type of equipment and labeling, sales can be handled rapidly at checkstands. The computer processes information as quickly as the checker picks up and moves the products across the reader.

The coding on the label of each supermarket package is a data item. The data from the labels are processed by the computer to provide several kinds of information:

- As far as you are concerned, the most important information produced by the computer is your grocery bill. This is a listing of the items you bought and the amount you owe. These data items and the information about your grocery bill appear on a printed tape. The same data and information are

**Customer receipts are computer-produced** in many modern supermarkets. Thus, many people encounter computers during their everyday lives.

displayed in lights at the top of the sales register. The processing of data items from packages delivers the information you need to pay your bill. The total amount you owe is information.

- When you present money to pay for your purchases, a new data item is created. This is the amount of money received by the checker. This amount is entered into the keyboard of the sales register. When this data item is received, the computer compares the amount you have paid with what you owe. If you have paid more than you owe, the computer figures your change. The change you are to receive provides information for you, and also for the checker. The management of the store also gets important information. The computer totals the amount of money received in each register, every day. This information indicates the value of items sold in the store. The computer also figures the total for sales at all checkstands. This is the amount of money, or cash income, that the store has received. Certainly, this is extremely important information to the managers and owners of the store.

- Sales reports are also prepared by the computer. These are based on entirely different data items from those used to prepare grocery bills and cash receipts reports. Sales reports are based on the data items that identify products. These are the codes read from the special labels on food packages. The computer keeps track of sales by product or item. On a regular basis, these data are turned into information—reports on how many units of each product are sold. This information lets store management know just what customers are buying. Since the idea of a self-service store is to provide products that customers want, this information can be valuable indeed.

- Supermarket computers also produce *inventory* reports. These consist of information items on how many units of each product are in stock. Inventory is the stock of products on hand. This information is carried in the computer's electronic storage devices. Information on sales totals for products is then deducted from the inventory amounts. Management continually has information about how much of each product is in stock. Combining information from the

Distribution of most products that people eat or use in other ways originates with computer processing.

inventory and sales reports, management knows how much of each product should be bought.

- Purchase orders are information documents that ask suppliers of products to ship goods to a customer. In buying inventory, the supermarket is the customer. The people who package food are the suppliers. By using information from inventory and sales reports to develop orders, the store is buying what its customers want. When computers are used, it is as though the customers are refilling the supermarket shelves with wanted products.

This everyday example demonstrates something about the nature of data and information. For one thing, data are processed to produce information. Another important factor is that information is *cumulative.* Cumulative means that information value is built as new items are added to previous results. Each of the reports described above tells store management something about what is happening. But, as new information is added, management gets a fuller picture.

**People require information** as a basis for performing their jobs and for making decisions. Computers have become vital decision-making tools.

Use of information, in turn, builds knowledge. As managers use information to make decisions, they are able to derive additional meaning from available information. This additional meaning is part of the knowledge that goes into running a business. Managers who pay attention to information on their operations, for example, become sensitive to customer needs. At different seasons, or as customer tastes change, products may be stocked in different locations within the store. Managers also learn—build knowledge about—what kinds of neighborhoods they should look for in building additional stores.

Thus, there is a continuing chain of development. This runs from data, to information, to knowledge. Data and information are processed and supplied by computers. Knowledge results largely from human intelligence. Remember this point: Computers are machines for use by people. The value of computers lies in the ability of people to direct the machines.

## THE INFORMATION EXPLOSION

Computers have been called agents for change. This statement seems to assume that computers bring about, or cause, change. This outlook on computers is not entirely true. Computers are tools for helping to implement changes. But this is different from being a cause for change.

As with other branches of technology, information processing techniques, including computers, have answered needs that people already had. This is quite a bit different from being a principal, or primary cause of change. People needed and wanted the power that computers can deliver. If this demand had not existed, computers never would have experienced the success they have enjoyed.

As with other forms of technology, then, a human need existed first. Computers provided an acceptable answer, at the right time. The need that led to acceptance of computers resulted from rapidly growing—some have called it exploding—volumes of information. The United States led the way in establishing new information demands—and in technology to meet those demands.

The so-called computer age took shape during a period that ran roughly from the mid-1930s through the mid-1950s. During the 1930s, bad times gripped the United States. The period was known as the Great Depression—a time when many people were out of work. The federal government passed many laws during these years that were aimed at helping people. The idea was to put people back to work—also to provide protection for the future security and retirement of working people.

### Complications Through Payroll Deductions

Beginning in 1936, most employers in the United States were required to begin taking social security deductions from workers' pay. The money went to the government, which guaranteed retirement and other benefits to the workers in exchange for the money paid through payroll deductions.

**The data processing explosion** started with the requirement for social security deductions introduced in 1936.

**Punched card processing** required the use of multiple machines. This machine was capable of reading data from one set of cards and copying those same data onto new cards. (Courtesy Sperry Univac)

When you think about a single paycheck, social security deductions may seem simple. The employer was required to take 1 percent of each worker's pay. This money was to be held and paid over to the government periodically. In addition, the employer was required to make similar contributions to the social security fund.

Before social security, companies simply paid workers what they had earned. If people received a salary, the same amount was paid out each week. If people were paid on an hourly basis, one simple calculation did the trick. That is, the employer simply had to multiply the pay rate by the hours worked. The result of this single multiplication produced the worker's earnings.

The requirement to withhold pay literally multiplied the data and information processing requirements for employers. Just to figure a paycheck, at least three times as many arithmetic calculations were necessary. In addition, however, employers had to set up separate records to account for social security contributions by individual workers. Further, still other records were needed to report to the government.

In total, the seemingly simple requirement to withhold social security taxes multiplied data and information processing requirements by 10 to 15 times. That is, it took between 10 and 15 times as many calculations and record keeping steps to handle payrolls. Data and information processing became a big business almost overnight.

## Punched Card Processing

During the 1930s, increasing demands for data processing were met by *electromechanical accounting machines.* These were devices that made it possible to capture data items once and to reuse them many times without extra pencil-and-paper work. The means of capturing data were *punched cards.* These were pieces of thin cardboard into which holes were punched. The positions of the holes indicated the values of the data. These cards could then be processed on electromechanical devices sometimes called *tabulating machines.* To tabulate is to list and total data items in an organized way. The tabulating machines got their name from the fact that they could read data from punched cards and produce tabulated reports. During the first years of the information age, emphasis was on volume production of bigger, faster tabulating machines.

## Wartime Buildup

Demands for information increased even more explosively between 1940 and 1946. These were the years in which the United States armed itself for and fought in World War II. During the war, some 10 million persons were taken into the Armed Forces. Massive amounts of information had to be processed to identify these people. Still more processing was needed to assign the personnel to military units. Of course, the Armed Forces also had its own payrolls to process.

Information processing also became more complex outside the Armed Forces. In 1943, laws required that taxes on workers' incomes be withheld by employers and turned over to the government. These demands occurred at a time when individual companies were growing larger than ever. Part of this growth was necessary to meet demands of the entire free world for weapons and other war materials.

**One of the first computers to process data electronically** was ENIAC, which stands for Electronic Numerical Integrator and Calculator. The equipment in this historic photo weighed 330 tons. (Courtesy Sperry Univac)

During these same years, the war itself was stimulating development of new technologies. Advanced systems of radio communication were developed to support military units on battlefields. As aircraft became major weapons of war, new defenses were created. One of the technologies that resulted was radar. Radar resulted from the development of the new technology of microwave transmission. A microwave is a short, precise type of radio signal. With radar, microwaves can be bounced off a distant target. The waves are then returned to the transmission point. These bounced waves can be used for precise measurements. Radar permitted the tracking of aircraft targets. The advanced radio and radar techniques of World War II led to great advancements in electronics.

**The first computer used commercially** was this UNIVAC I. (Courtesy Sperry Univac)

## Post-War Advances

By the end of World War II, advances in electronics were ready to deliver television to the American public. The same electronics industries were also ready to produce computers for business and government use.

The first computer put into use in the United States was purchased in 1948 by the U.S. Bureau of Census. This computer was used to count and report on the census of 1950. It seemed fitting that computers should get their first use in the Census Bureau. Tabulating machines had also gotten their start in census calculation. The Bureau of Census had used the first, crude tabulating machines on the census of 1890. These machines had been invented by a Bureau of Census employee, Dr. Herman Hollerith.

Many challenges had to be met before computers were accepted as business tools. The computer did prove its value in the processing of the 1950 census. At that time, however, it was still believed that computers were useful only for massive data and information processing jobs. Computers were not believed to have a place in everyday business processing. At the time, one of the leading executives in the data processing industry predicted a limited computer market. This executive believed that the whole world would be able to use a total of only 300 computers.

Things didn't work out that way. By 1980 there were already millions of computers in use. Computers had taken on many shapes and sizes. At the small end, computers were used at home by hobbyists. At the larger end, computers were monitoring space exploration and advanced scientific experiments. In between, the great majority of businesses in the United States—and in many other countries—had become computer users.

Once the demand had been proven, technology helped to make computers ever more affordable. During each five-year period since approximately 1950, the cost of computing power has been reduced by approximately 90 percent. That is, in 1955, processing and memory devices cost about 10 percent of what they did in 1950. This pattern was repeated in 1960, 1965, and through into the 1980s. These cost reductions have resulted from advances in technology. Today's desk-top computers have more power than the room-sized computers of the early 1960s. These technological advances are dramatic.

**Computing power** can come in tiny packages. This hand-held calculator is designed to perform real estate calculations. (Photo, courtesy of Hewlett-Packard Company)

**Many modern jobs** require the knowledge and skills necessary to use computers.

Just as dramatic has been the growth in information and knowledge requirements that have been met by these new technologies. Keep in mind that computers are tools. If organizations and individual people did not create increasing demands for information processing, the technology, by itself, would have gone nowhere.

## THE INFORMATION PROCESSING CYCLE

Most computers in use today process data and deliver information. Therefore, it will help you to understand what information processing is and how information processing is done.

Information processing involves principles that are older than and separate from the use of computers. Computers are modern tools that have helped improve data processing methods and abilities. However, the principles of data processing are separate from the principles of computer operation.

People were the first data processors. People processed data and developed information for centuries before computers were introduced. People are still the key necessities to achieve data processing and information development.

Whether or not computers are involved, a basic, long-established *information processing cycle* is followed. This information processing cycle consists of four parts. These four parts are present in any system, or set of procedures, that handles data and delivers information. These parts are:

- Input
- Processing
- Output
- Storage.

### Input

*Input* encompasses all of the activities and paperwork connected with identifying data and introducing them to the information processing cycle. In business, most data are created as part of *transactions.* The transaction is the basic act of doing business. A transaction occurs whenever something of value is exchanged. The exchange can be between one business and another or between a business and a person. For example, when a worker finishes a week on the job, a time card is turned into the payroll department. This time card is a transaction record. The worker has exchanged a value, his or her time, for another value, a paycheck.

**People are essential** to the processing of data and the development of information. This is true whether computers are used as tools or whether the work is done with human brains. People were the first data processors.

The act of buying or selling goods and services also leads to transactions. Transactions also occur when businesses report payroll or tax information to the government. Large businesses may take part in literally millions of transactions each week. Each transaction, in turn, represents data input to the organization's information processing cycle.

When the input step is completed, data are *recorded* in some usable form. The data are then ready for processing. If the work is done by hand, input may be accomplished by writing numbers on paper. Computer procedures for handling input are discussed in the chapter that follows.

The same is true for computer operations in other parts of the information processing cycle.

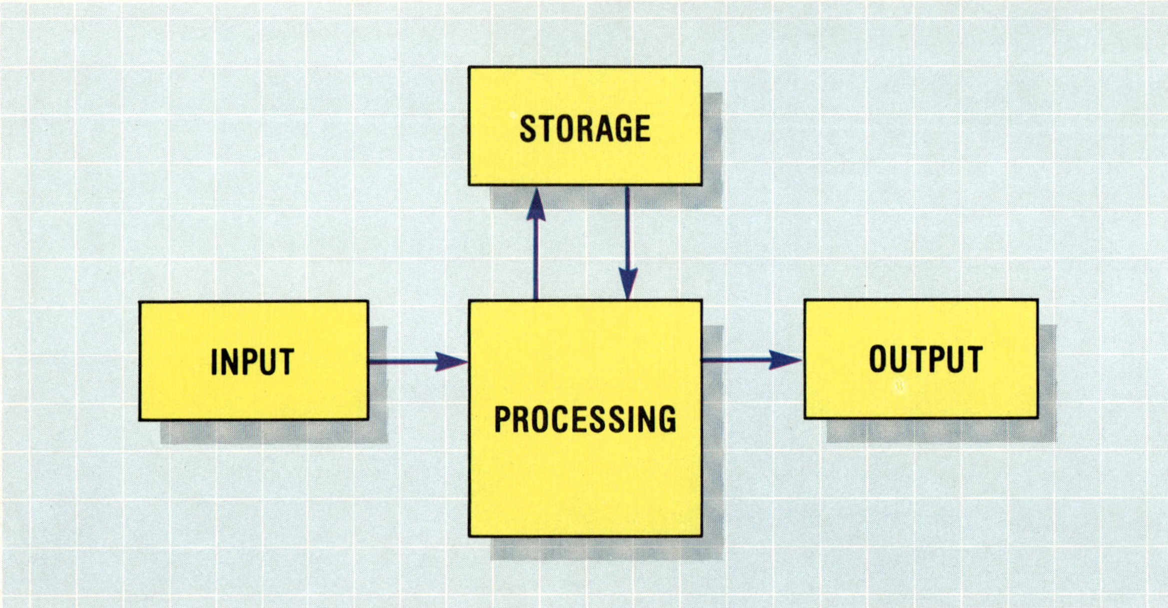


**The information processing cycle** can be represented by this diagram. The parts of the process are input, processing, output, and storage.

## Processing

*Processing* is the handling of data and the *transposition* of data into information. The word transposition means that data are changed by the processing that takes place. Transposition of data results when arithmetic functions are applied. When two numbers are added, their sum represents a transposition from data to information. The same is true when arithmetic functions are applied to data that have been input.

Transposition can also result from choices among data items. For example, suppose a business has several prices that can be charged for the items it sells. The different prices are charged to different categories, or classes, of customers. The input portion of a system, then, would have to *classify* each customer. Based on this classification, the proper price would be selected. For example, computers sold to schools usually carry a lower price than that for the same unit sold to private business. Thus, an information system to process computer orders might have a step that classifies customers as schools or private businesses. From this classification, the proper price can be applied.

Another type of processing involves comparison to identify what kind of customer is placing the order. Also involved is selection of the proper price to be applied. These functions—*arithmetic, comparison,* and *selection*—are the basic processing capabilities of information systems. Thus, information processing is not complicated. Rather, information processing involves the imaginative use of a series of extremely simple steps.

## Output

*Output* describes any method for delivering information to those who need it—the users. Output can be delivered as printed reports, as business transaction documents, such as checks or bills, or as displays on TV-like screens. Output can also be as simple as a telephone call that answers a user's question. Any method that delivers information in usable form qualifies as output.

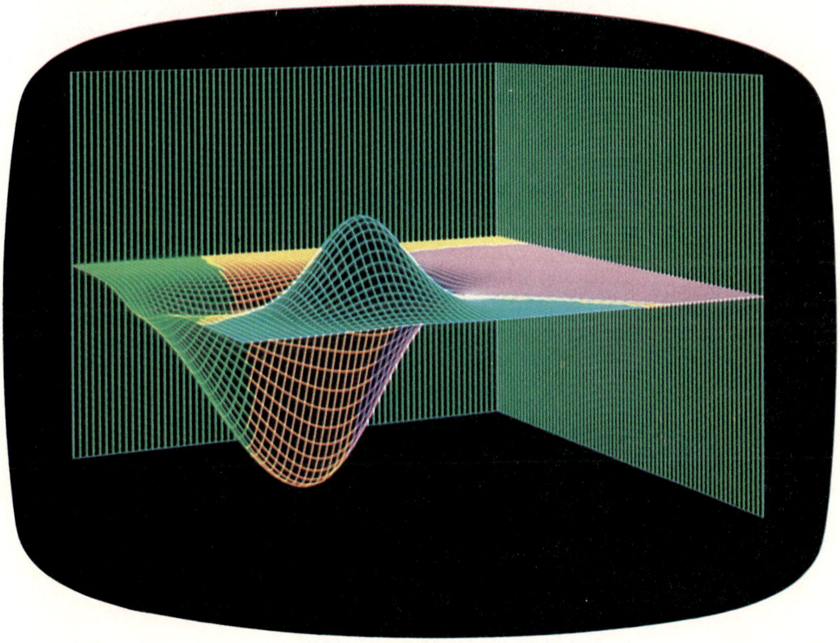

**Pictorial information** is also developed by computers for use by people. Many products are now designed and built with computer graphics capabilities. (Courtesy Los Alamos National Laboratory)

## Storage

*Files* of data and information are vital resources for any company. Data input to and information developed by processing systems become necessary for companies. Most companies could not do business if they did not have files of data and information. The term file refers to a collection of data and/or information that is organized in usable form. That is, the items of data or the collections of information are related to one another.

**Data media** include any methods or materials usable for storage and reuse of data. Included are magnetic media such as computer tapes, as well as ordinary paper documents and file folders.

For example, collections of information about customers may be collected according to customer name. Information on all customers would be part of a file. Information on a single customer, then, would be known as a *record*. The organization plan for a customer file might be *alphabetic*. That is, the records are organized alphabetically, according to customer name.

If numbers are assigned to customers, the organization may be *numeric*. Records would be placed in the file in order, from the lowest numbers to the highest.

Both alphabetic and numeric organization plans produce *relative* file organizations. That is, all of the records in an alphabetic file would be relative in position to the first record under the letter A. In a numeric file all of the records would be relative to the first record with the lowest number.

Within information processing systems, *storage* is carried out through some technique for building records and accumulating files of records. If processing is done by hand, the records can be pieces of paper accumulated in file folders. The file folders become records. The total group of file folders becomes the file used to support the system. If computers are used, records and files are put together electronically and automatically.

It is impossible for the information processing cycle to be carried out without some method for storing records and files. Remember, information processing takes place in cycles. Cycle means that things happen again and again. In business, transactions lead to other transactions. When a company sells something, an invoice (bill) is sent to the purchaser. A record must then be set up to collect that bill. When the bill is paid, those records must be available for further processing. Just as a company operates continuously, its files must also be usable continuously. That's why data and information are called vital resources for information processing.

Remember, these are the basic principles of information processing. They apply to any *information processing system.* Some method of storage is needed if data are processed by hand or by computer. These needs don't change.

The parts of the information processing cycle, as indicated, make up an information processing system. *System* refers to a series of parts that work together to perform certain functions and to deliver expected results. The use of computers requires the building of systems, as described in the chapter that follows.

## Summary

The relationships among data, information, and knowledge are important to an understanding of computer technology. Briefly, computers accept data, produce information, and people use the information to build knowledge. Data are basic facts. Information is formed by relating data items. Knowledge comes from the use and analysis of information.

The computer is a set of devices that can process data in a complete sequence of steps—uninterrupted. A calculator can also process data, but it requires a human operator at each step. By contrast, the computer can follow a program to complete the job on its own.

Computers produce information quickly and efficiently. This information can be stored and accumulated. The more information people have about a topic, the more their knowledge can be increased. The United States was first to develop potential uses of computers. This is why it is said that the United States has become an information society.

Computer technology is developing in the same pattern as other technologies. First, a need existed. Technology met the need. Now that the technology is available, people have developed new needs.

When the social security system was first established, people in government needed a more efficient way to handle data and information. World War II also brought a need for better methods of record keeping. People saw a need. By applying existing technologies and developing new ones, the computer was created. The first electronic, programmable computer was used by the U.S. Bureau of Census in 1948.

Whether using the human brain or a computer, the principles of information processing remain the same. First, data are collected and recorded. Next, data are changed to information by performing certain functions, such as applying arithmetic. Next, information is delivered to someone who needs it. Finally, information is stored.

As with any technology, people involved with computers have developed a special language. For instance, the first step of information processing is to collect and record data. In computer language, this is called input. The second step, changing data to information, is called processing. Delivering information to those who need it, or users, is called output. Saving information so that it can be used again is called storage.

## Technology Terms

- information society
- data
- information
- knowledge
- data item
- processing
- computer
- program
- bar code
- inventory
- cumulative
- electromechanical accounting machine
- punched card
- tabulating machine
- information processing cycle

## Using Technology

1. Identify which step of an information processing system is being performed in each of the following activities.

   - A student jots down notes while listening to a class lecture.
   - A secretary delivers a typed report to the boss.
   - A waiter adds up a customer's bill.
   - A cook saves a favorite recipe in a recipe box.

2. Directly or indirectly, computers affect everyone. How many situations can you think of where computers serve you? Bring your list to class for discussion.

3. Why do you think the United States has been called "the information society"? Write your answer in a short composition.

input
transaction
record
process
transposition
classify
arithmetic
comparison
selection
output
file
record
alphabetic
numeric
relative
storage
information processing system
system

# 13 COMPUTER SYSTEMS

### Your Learning Job

After you have read this chapter and completed the work assignments, you should be able to:

- ☐ List and describe the parts of an information processing system.
- ☐ Explain the roles of people in computer information systems.
- ☐ Describe two types of procedures that every computer information system must have.
- ☐ Describe several types of data media.
- ☐ Explain what the central processing unit of a computer does.
- ☐ Describe several forms of output.
- ☐ Explain the difference between primary storage and secondary storage in a computer information system.

## COMPUTERS AND INFORMATION PROCESSING SYSTEMS

A *computer* is a collection of devices that function together for automatic, continuous processing of data. To process data continuously, computers need sets of instructions that can be stored internally. These sets of instructions are known as *programs*. Sometimes the collections of devices and programs are referred to as a *computer system*.

**Computer systems** are designed and used by people. People, in turn, require extensive documentation, or paperwork, to guide job performance.

However, neither a computer nor a computer system by itself can do the complete job of information processing. A larger entity, an information processing system, is needed. Most modern information processing systems use computers. However, computers are just one of a group of parts that make up an information system. These parts of, or requirements for, information processing systems include:

- People
- Data
- Procedures
- Processing equipment, including computers.

**Complete libraries** of data and information are generated by large computer systems. (© Liane Enkelis)

## People

People are the first requirement for any data processing system. Without the critical functions performed by people, there would be no information processing systems using computers. The roles of people in computer information systems include:

- People are the customers for computer information systems. Remember, all systems of technology meet demands by people. Thus, people who need information support the technology for development of computers and of information systems.
- People design information systems and specify what they are to do. High-level experts, called *systems analysts*, work with information users. Systems analysts determine what user needs will be. Then the analysts figure out whether the requirements can be met through use of computers. If so, the analysts specify what equipment is to be used and what processing should take place.
- People design and write the programs containing instructions computers will follow. Without these programs, computers are helpless collections of wires, cabinets, and electronic parts. The processing performed is only as good as the quality of programs prepared by people.
- People set up and operate the computer equipment. Other people maintain the equipment. The files required to run jobs on computers are also maintained by people. All of these functions come under the heading of *computer operations*.
- People must originate the data that serve as computer input. Many thousands of *computer input operators* use special equipment to present data to computers in usable form.
- People monitor the quality of the work done by computers. These assignments are handled by a number of specialists.

**Cooperation among people** is a key to the design and development of computer systems. (© Liane Enkelis)

Their job is to make sure that data entered into the computer and information delivered are accurate, reliable, and produced on time. Without assured quality, information systems have no value. The people who are trained to monitor computer operations and provide assurance of quality, therefore, play key rolcs.

## Data

Data are the raw materials of information systems. Without data to process, there could be no information to deliver. Data are just as much of an essential raw material for information technology as are the raw materials for other technologies. For example, it is necessary to grow wheat so that flour can be ground and

**Storehouses of data** often develop around large computer systems. For protection, computer tape files of this type are usually maintained in fireproof vaults.
(Four By Five, Inc.)

bread can be baked. Similarly, it is necessary to mine iron and coal so that these raw materials can be processed to produce steel.

In some descriptions of computers and information processing systems, the importance of data is overlooked. In studying and understanding information technology, you should never lose sight of the critical importance of data. Development of information systems requires careful study and planning to create and process data, and to transform them into useful information.

## Procedures

All information systems need procedures. The completion of procedures leads to the transformation of data into information. Procedures also guide the delivery of information to users. To make an information system work, all data items must be in the proper, assigned positions within every record. Every record, in turn, must be in a proper position within its assigned file. This control of data and information is essential to the delivery of usable output.

Think of your own experiences with information system inputs and outputs. You follow standard input procedures whenever you fill out a form to be used in an information system. Examples include registering for school, signing up for

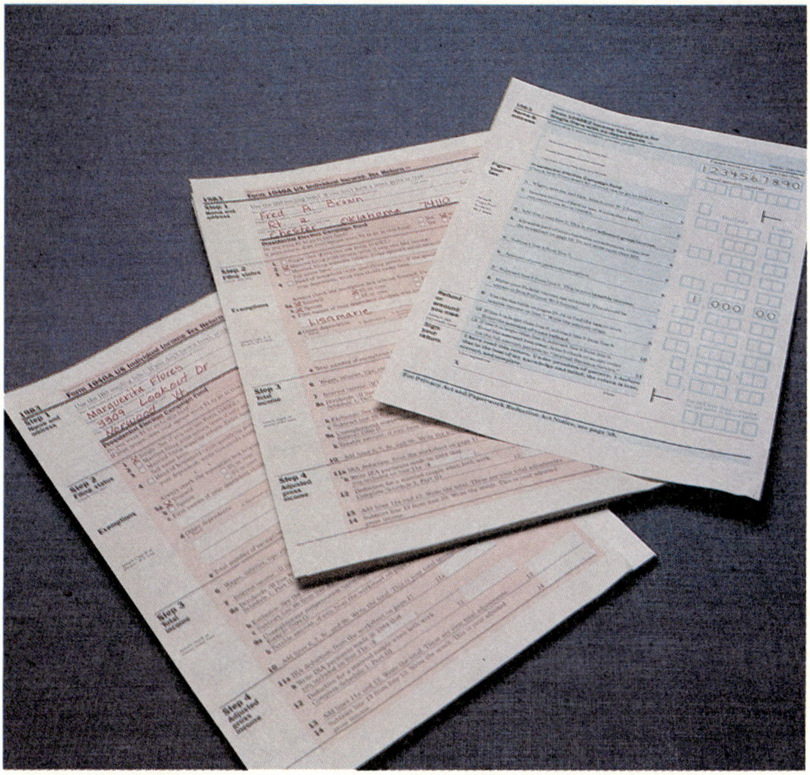

**Uniformity of data** is vital for the design and use of computer systems. Uniformity is necessary because government agencies establish specific standards that all businesses must meet.

a library card, or applying for a social security card. When you sign up for activities such as clubs or Little League, there are also procedures to follow. If you have your own bank account, you have also had to fill out account and signature cards.

All of these forms ask for specific data items in assigned spaces. The form you fill out is exactly like thousands, or millions, of others. You have been following procedures that were developed by the designers of the system. *Uniformity* in the recording and processing of data are critically important to the success of an information system. Uniformity means a sameness, or regularity, among the parts of a system. In systems that are operated by hand, uniformity makes things easier for people. If a computer becomes involved, uniformity is essential. Without uniformity of data-item organization and processing procedures, computer systems could not exist. Remember, programs written by people are stored within computers. Computers follow these programs reliably and repeatedly—time after time. Any data or processing requirements that are not uniform will be rejected by the computer.

**Procedures for computer systems** are usually documented in large manuals. Part of the skill of computer professionals lies in following specific instructions.

(Photo courtesy of Hewlett-Packard Company)

To understand the value of uniformity, think of some of the information system outputs that you know about. One example is your school report card. Each subject you study has its own place on the report card. Grades must be reported according to subjects. It would not do to mix up a bunch of grades and a bunch of subjects. You would have no idea of how well you were doing, in which classes. The very purpose of putting out report cards would be defeated.

Think also about what would happen if uniformity was destroyed in a company's payroll data. An information system that prepares payrolls must have uniformity. This is necessary because each worker's time on the job or salary must be placed in the proper employee record. Without uniform handling and accurate data, there could be no modern payroll systems.

Each information system that uses computers needs at least two kinds of procedures:

- Job instructions
- Programs.

**Job instructions.** Job instructions cover work to be done by people. Written instructions are needed for every information system. Even if the system uses computers, people still play key roles. People are responsible for computer operations during each stage of an information processing cycle.

Job instructions provided for people are usually called *manuals.* A manual is a document that provides job instructions or explains the use of equipment. An information system manual has specific job descriptions for each task to be performed by people. For example, an information system manual might tell one employee which files are to be used in processing a given job. Another manual might contain instructions for delivering system outputs.

Manuals used for job instructions in information systems are similar to those provided with many products. For example, every new automobile has an owner's manual. If you buy a new camera, radio, or stereo, you will receive printed instructions with the equipment. These, in effect, are manuals telling how to use and care for your purchases.

**Programs.** Programs, in effect, are manuals in the computer's own language. Programs direct the use of computer equipment.

**Computer-produced documents** often play important roles in the continuing use of information systems. Computer programs are commonly printed to create documents like those shown here. (© Liane Enkelis)

In computer technology terms, the computer equipment, collectively, is known as *hardware*. The programs that direct use of the equipment, in turn, are called *software*. The term hardware describes a piece of equipment or a series of connected devices. The capabilities of hardware are established when equipment is manufactured. Software, on the other hand, is flexible. People can change software. Software applies the capabilities built into hardware. Software directs hardware in performing specific jobs. There are two broad types of computer software:

- System software
- Application software.

System software is usually provided by the equipment manufacturer. These software units consist of special programs that direct the operation of equipment. To use a computer, for example, it is necessary to perform a continuous string of input, processing, output, and storage operations. A number of separate, connected hardware devices are used to do this work. The system software serves to link the devices and to direct their coordinated use.

Application software controls the processing of data for individual jobs. An application program is written by people. Application programs direct specific operations needed to

**People use computers** through keyboards and display screens. Increasing numbers of people encounter computers in their everyday jobs. (© Liane Enkelis)

transform data into information and to deliver results. These programs are written in special *programming languages.* A programming language consists of a set of special terms, or instructions, that can be understood by both people and computers. Each programming language also has a special translator program, known as a compiler, that is housed in the computer. When the instructions written by people are entered, the compiler translates them into *machine language.*

Machine language describes the set of special signals that operate the computer. Remember, computers are electronic. Therefore, computers are controlled by electrical circuits. The controls applied within computers are basically the same as those that you use to turn lights on or off. This means that the language of computers is limited to signals with two choices. This signaling system is known as a *binary* language. Binary means two. A binary language operates in sets of on or off values.

System software programs use binary language to control the computer. The compiler language, in turn, creates binary instructions. People in the United States use a communication system that includes 26 letters in an alphabet and 10 numerals. There are other, special symbols, such as punctuation marks. Compiler programs accept instructions in the language of people and translate them to the language of computers.

There are literally hundreds of programming languages available. Some popular languages include FORTRAN, COBOL, Pascal, PL/1, RPG, and many others. The BASIC language is most commonly used with small, personal computers. Also, BASIC was developed specifically to teach students how to program computers. Therefore, it is in wide use. An accompanying chart shows some of the people-related terms of the BASIC language. Also shown are the accompanying computer operations that result when these instructions are processed.

## Equipment

A computer is a series of devices that are connected to work closely together to process data and deliver information. Under control of programs, computers can run through complete processing cycles without help from people.

| BASIC Statement Type | Usage |
|---|---|
| CLOSE | To disassociate a data file from a program |
| DATA | To define files that are used as constants within a program |
| DIM | To define and reserve areas within memory to be used as program tables |
| END | To identify the physical end of a program |
| GOSUB | To transfer control to a program subroutine |
| GOTO | To transfer control from one point in a program to another |
| IF | To test a data field for a particular value |
| INPUT | To accept data from an input unit and to retain values in memory |
| LET | To assign a value to an area in memory |
| OPEN | To link a data file to a program |
| PRINT | To transfer data from memory to an output unit |
| READ | To input data from a file of constant data in memory |
| REM | To provide documentation for a program |
| RETURN | To transfer control from a subroutine back to the calling module |
| STOP | To cause the computer to halt processing |

**Comparatively few instructions,** when used in the right order, can cause a computer to perform complex processing.

The term computer refers to any set of equipment that can store and follow instructions of programs. The functions performed by computers, as you know, include input, processing, output, and storage. Special units of equipment are used for these functions, though some units serve two or more functions.

The single word, computer, can be used to describe many thousands of different products that come in a wide variety of sizes. A computer can be small enough to fit into a briefcase or on a desk. Computers can also be large enough to occupy complete buildings. Prices of computers can range from a few

**Computers seem to get smaller** all the time. Many manufacturers now make portable computers. (Courtesy Radio Shack, Division of Tandy Corporation)

hundred dollars to tens of millions of dollars. However, to be called a computer, any collection of equipment needs to be able to do two basic things. A computer must store and follow programs and must perform the basic functions of input, processing, output, and storage.

The design and construction of computer equipment is related closely to the functions to be performed. That is, devices used for input, processing, output, and storage all have special features. These features accommodate the work done by the devices and the people who use them. Therefore, specific units of computer equipment are described in the section that follows. Also, the discussions cover the work of people who use the equipment.

## COMPUTER OPERATIONS

The functions of computer equipment and the efforts of people who operate computer systems are related closely. The natural way to describe these operations is in terms of the functions performed:

- Input
- Processing
- Output
- Storage.

### Input

The method used for data input depends on how and where data are created. Basically, most data used in information systems originate in *source transactions.* A source transaction, as described earlier, is any basic act of doing business. Included are transactions such as the buying or selling of products or paying of workers. To complete a transaction, it is usually necessary to prepare a *source document.* A source document is a record of the data produced during a source transaction. Examples of source documents include checks, invoices (bills), or purchase orders.

Three basic procedures can be used in connection with data input. These procedures can be used separately or together. They are:

- *Data capture* consists of any method for encoding data in a form usable by computers. The function of data is related closely to the source transaction. Data capture can involve *transcription* of data from source documents into computer-usable form. Most typically, data from source documents are

**Data capture and data entry** can often be combined into a single input operation. Using terminals like the one shown here, operators can both capture data and complete entry into computers in a single operation.

read by people and entered into devices using keyboards similar to those of typewriters. These keyboard-operated units may be used to enter data directly into computers. Just as often, the data are recorded, or captured, on special *media*, or data media. These media can be documents that can be read into computers or they can be storage media. Storage media are described later in this chapter. No matter what system is used, the function of data capture involves the transition of source data from human-readable to computer-readable formats. Captured data are ready for computer input and processing. This means that the data entry function can be broken into two parts. They can be separate steps for the capturing of data and the actual entry of data into computers.

- *Data entry* is the actual presentation of data to a computer for processing. Data entry includes both the reading in of data and the setting up of those data in preparation for processing. Data entry can be performed either by people working at keyboard-operated devices or by automatic machines. These methods are described in the next paragraph.

**Source transactions** are frequently entered directly into computers through use of electronic sales registers like this one.

- *Direct input* means that data are both captured and entered into a computer as part of the source transaction. Recall the example of the supermarket checkstand. The data to be entered are captured in advance, on package labels. Data entry takes place as the checker moves a package across a scanning device in the checkstand. No transcription step is necessary—except for special products such as meat or vegetables. As the computer reads data from labels, transcription and entry functions are performed automatically. This is one popular method of direct input. Direct input can also take place from keyboards. For example, many banks now use *automatic teller machines.* These are computer-connected devices that are usually set up in bank lobbies or outside of banks. Some automatic teller machines are set up in shopping centers or other locations. Each automatic teller machine (ATM) is part of an overall computer equipment system. Direct entries can be made from the ATM into the computer. These entries, however, are not fully automatic. Rather, the customer who wants to deposit or withdraw money must operate a keyboard on the ATM. Thus, the capturing portion of the direct input function can be either automatic or through transcription (keyboarding). The result remains the same: Data created as part of source transactions are both captured and entered in a single operation.

**Data capturing equipment.** Data capture is the recording of data on some medium that can be read automatically by computer-controlled devices. Most data capturing units are equipped with keyboards similar to those of typewriters. A data medium consists of any material on which data can be recorded in computer-readable form. There are many, many types of data media. However, the most common media include:

**Data media** include computer tapes, magnetic diskettes, and punched cards. Plain paper and file folders can also store data for system use.

- Punched cards
- Magnetic tape
- Magnetic disk
- Magnetic diskette
- Magnetic cassette
- Specially encoded documents.

The main method of capturing data for recording on all of these media is *keystroking*. The term keystroking, simply, describes

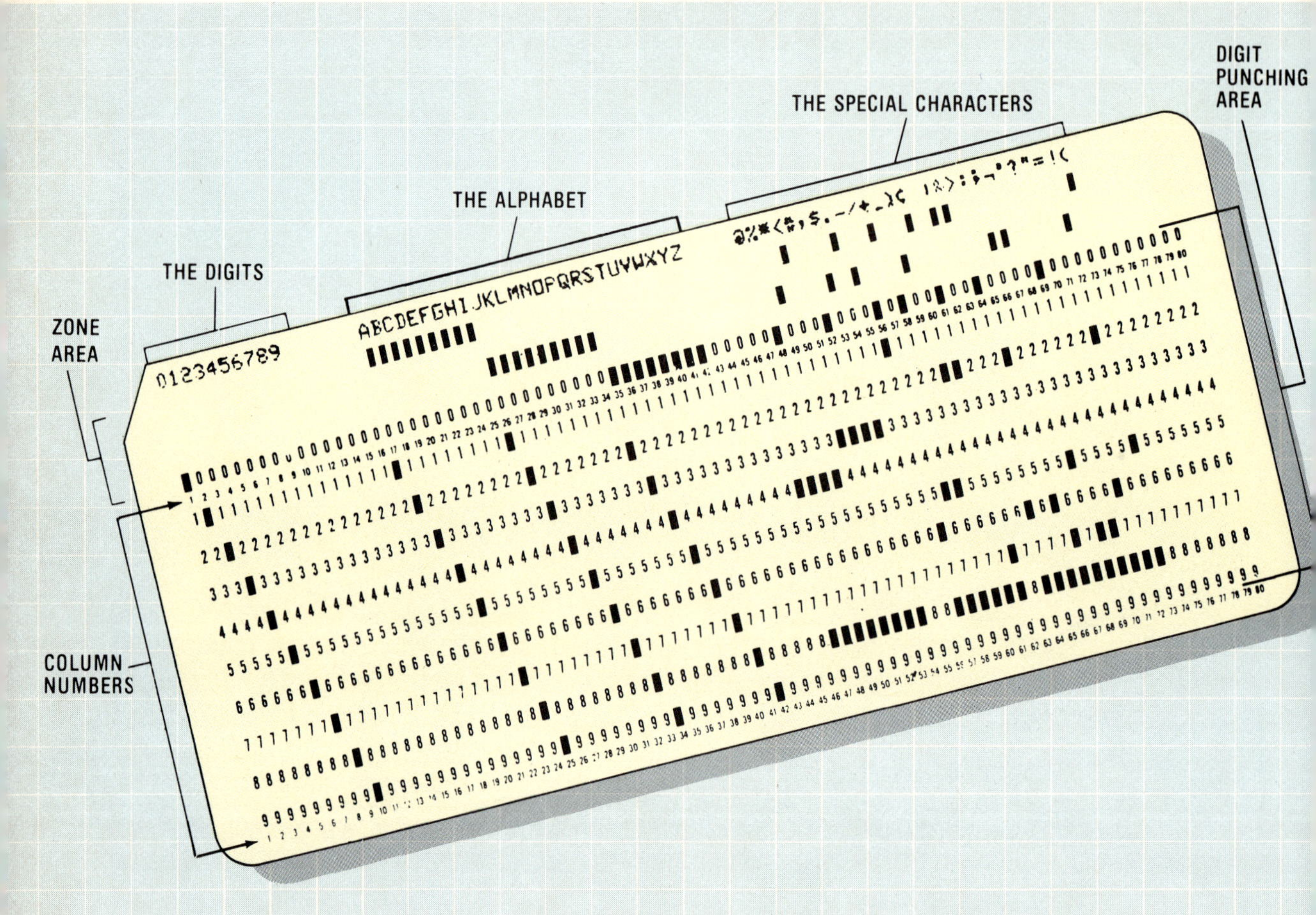


**The punched card** was the original medium for recording of data in modern data processing systems. Punched cards are still in use in some systems.

any operation involving keyboard entry. Procedures for entering data through the touching of keys are all grouped under the heading of keystroking.

*Punched cards* were the original media of modern data processing. Punched cards were used by the original tabulating machines for processing the 1890 census. By the 1950s, many millions of punched cards were being used every day. To highlight the volumes involved, consider that the cards are made from the same material as milk cartons. During the 1950s, more of this material went into punched cards than into milk cartons. From the 1960s onward, use of punched cards declined. Newer machines use magnetic media and specially encoded documents, as described below. The illustration above shows the format of the punched card, including the coding used for data.

*Magnetic tape* consists of a continuous ribbon of acetate (plastic) material that has been specially coated. The coating on magnetic tape consists of iron oxide. This material can be magnetized. In effect, positions on magnetic tape are magnetized in much the same way as holes are punched in cards. The difference lies in the speed and capacity of magnetic tape. A single inch of magnetic tape can record up to 6,250 characters of data. By contrast, a single card measuring more than seven inches in width records only 80 columns of data. Tapes can also be read and processed faster on computers than cards.

An accompanying illustration shows a typical format for the recording of data on magnetic tape. Note that there are nine recording positions, known as rows or *tracks*. Data characters are recorded in *vertical columns* along the length of the tape. The

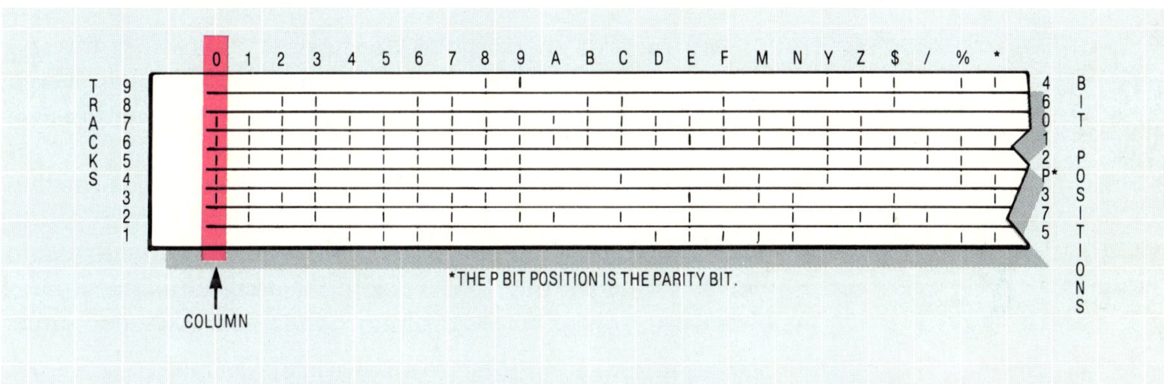


format shown has nine data, or *bit* positions. A bit is a *binary digit*. This means that each bit has a value, for computer processing purposes, of 0 or 1. Eight of the bits in the nine-bit format have data values. The ninth position is set aside for a *parity bit*. This parity bit is used in checking accuracy of the recording and reading of data by the computer.

*Magnetic disk* media resemble phonograph record albums. However, recording is done magnetically on disks used by computers. Each disk usually consists of a metal base coated with the same type of iron oxide used for magnetic tape. Data are recorded in circular patterns, also called tracks, around a central hub of a disk. A read/write mechanism is positioned immediately above the recording surface of each disk. On instructions from the computer, the read/write mechanism finds the proper

**Magnetic tape** is a popular data medium because it provides compact data storage. Tapes can also be recorded and read at high speeds. However, tapes must be processed sequentially, causing problems for some applications.

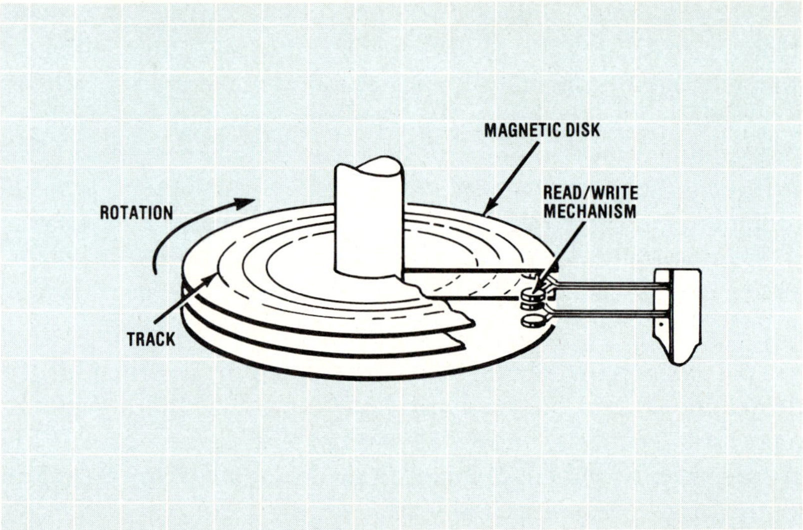


**Magnetic disk** media permit direct access to data and information. The photo above shows the changing of a disk pack. The drawing at right shows how data are recorded on disk packs. (© Liane Enkelis)

track and records or *retrieves* the needed data. In computer terms, retrieve means that a record is both located and delivered to the computer for processing. In the accompanying illustration, the read/write mechanism moves across the surface of magnetic disks. Some disk devices have a read/write mechanism, or head, above each track. Head-per-track devices are faster than movable-head devices. This is because the heads do not have to move to the right track to locate records. However, head-per-track disk devices cost more than movable-head units.

Data recording patterns on magnetic disks are different from those for tapes. Bits of data are recorded alongside each other on disks. This is different from the vertical pattern on tapes. Disk recording is *serial*. Serial means that all data bits are recorded alongside one another around a track. Disk recording patterns are shown in an accompanying illustration.

Disk media are of two general types—fixed and removable. The term *fixed disk* means that a series of disk platters is mounted permanently within a device known as a *disk drive*. Most fixed disk devices also have fixed heads. This means that they are relatively high in capacity and speed.

*Removable disk* media can be taken off their disk drives and stored separately. This makes it possible to use individual disk drives for handling multiple files. Disk media can be placed on disk drives when specific application programs are being processed. At other times, disks can be stored separately from the

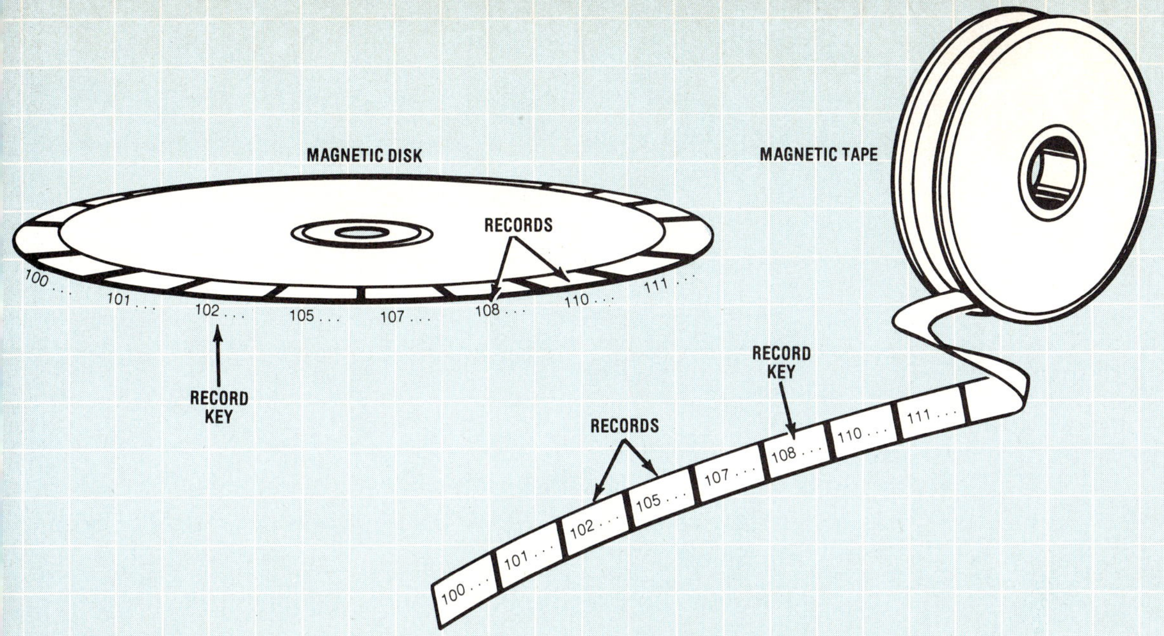


computer equipment. This means that whole collections, or libraries, of disk storage media can be built. Total storage capacity increases when removable disks are used. However, operation of drives with removable disks is relatively slow when compared with those that have fixed disks.

**The recording of data records** on both magnetic disk and magnetic tape media is illustrated in the drawing above.

Disk devices illustrate an important point about computer technology. The first disk storage unit was introduced in 1957. At that time, the storage capacity of disk packs was 3.5 million characters of data. Today, individual, removable disk packs contain well over 1 billion characters of data. Yet, costs for disk drives are in line today with what they were in the early 1960s.

*Magnetic diskettes* provide a special form of disk storage. Diskettes consist of individual, round sheets of oxide-coated acetate. Data are recorded on diskettes in the same way as described for disks. The main difference is that diskettes are small and inexpensive. Diskettes are used largely with personal computers. They are also used on many special data capturing machines. That is, as keystrokes are entered into these machines, the data are recorded on diskettes. The diskette-recorded data can then be read into the computer. On most personal computers, as described below, diskettes provide the main means of secondary storage.

Computers are able to record data on *magnetic cassettes* similar to those used for stereo or voice recording equipment.

**Magnetic ink character recognition, or MICR,** encoding is used on checks and other banking documents.

In general, these cassettes are the same size and look very much alike. However, the tape itself has to be of special quality to record data.

A wide variety of specially encoded documents can now be used for data capture. One example already given is the universal product code (UPC) used in supermarkets. With this method, data are captured from imprints on package labels.

Two other widely used methods for special encoding of documents include *optical character recognition (OCR)* and *magnetic ink character recognition (MICR)*.

ISBN: 0-538-16250-3

**Optical character recognition, or OCR,** encoding is used for credit cards or other documents requiring direct computer reading. Complete pages of documents can also be recorded in OCR.

Optical character recognition uses specially formed letters and numbers that can be read electronically by computer devices. Possibly the best example of OCR is the imprinting and processing of sales checks for credit card purchases. When credit card sales tickets are processed, transaction data are captured in machine-readable codes. The tickets themselves can then be fed into computers at high speeds. This means that people who are selling or buying products are actually completing data capture. Obviously, this saves work because computer professionals are not required to capture these data.

MICR is a special form of coding used by the banking industry. A series of numbers and symbols is imprinted along the bottoms of checks or deposit slips. The ink used for MICR characters contains iron particles. When processed by special machines, these imprints are magnetized. The magnetized numbers and symbols each give off a special code. Thus, the computer can process data directly from the imprinting on checks or deposit slips.

The descriptions above cover the media produced when data are captured. The actual work of data capturing is usually performed on special machines that have typewriter-like keyboards. These machines can be used either at the point of transaction or in special sections of a computer department.

**Direct input.** In *direct input,* the functions of data capture and data entry are combined. Data are both captured and read into the computer in a single operation. Direct entry usually takes place as part of a basic source transaction. Often, data entry is performed by a customer of the computer-using company.

The supermarket checkstand that reads package labels is an excellent example of direct data entry. Many sales registers at fast-food restaurants or department stores also use direct data entry. That is, data are both captured and entered as entries are made on keyboards of sales registers.

An example in which customers do their own direct entry can be seen in the case of automatic teller machines. The customer at an automatic teller machine uses a direct-entry terminal. The customer uses a keyboard to identify the transactions to be processed. That is, the customer indicates whether a transaction is to be a deposit or withdrawal. Amounts to be handled are also entered by the customer. In effect, a bank using a computer has no costs for input operations. Customers replace the input operation function.

**Automatic teller machines** represent a popular method for direct input. With devices of this type, customers transact their business directly with computers.

## Processing

The *processor* is the device in which actual computing takes place. Every computer system, from the smallest to the largest, has a processor. Some newly designed computers have more than one processor. However, almost all computers in use at this time have a single processor. The processor may be a large device housed in its own cabinet. Or, a processor may consist of a few electronic chips mounted on a single circuit board within a personal computer. In any case, the same basic structure exists and the same basic operations are performed. Each computer processor has two basic sections, or parts:

- The central processing unit
- Memory.

**Central Processing Unit.** The *central processing unit* actually operates the computer and also performs all arithmetic and comparison functions. The central processing unit has two parts.

One is the *control unit.* This device operates under instructions from system software. The control unit directs the operation of the entire computer system, including the devices that handle input, output, and storage.

The other device within the central processing unit is called the *arithmetic/logic unit (ALU).* The ALU performs the basic arithmetic functions, including addition, subtraction, multiplication, and division. The ALU also has the ability to compare two

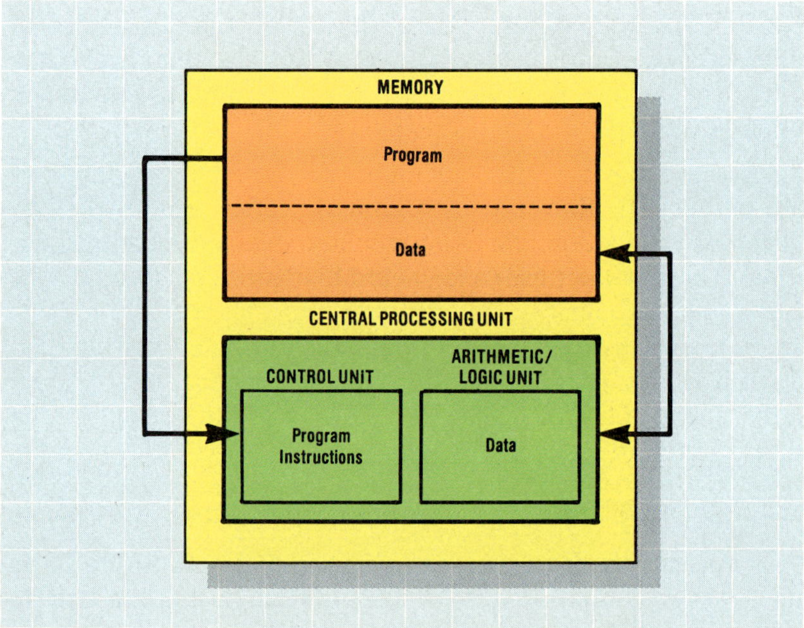


**A computer processor** contains a central processing unit and a memory. These parts of the processor interact in the handling of data.

data items. The ALU can then direct processing based on the results of such comparisons. The compared data items may be greater than, less than, or equal to each other. These three basic comparisons and the four arithmetic functions represent the total sum of computing capabilities. Any computer, no matter how large or small, does nothing more. However, computers can perform these functions at great speeds—in billionths of seconds. The challenge lies in planning for and organizing use of these basic functions.

**Memory.** The memory within a processor is also called the *main memory.* This device uses high-speed electronic circuits. High speed is necessary because the main memory must work at the same speed as the computer processor.

The main memory holds just those data and programs that are being processed at any given moment. Main memory is described as dynamic. That is, the programs on data in memory change continuously. These changes restructure the content of memory to meet the needs of programs being processed.

An accompanying diagram shows the general organization and operation of a computer processor.

Although operation of a processor is highly automatic, people still play important roles. People interact with a computer processor. The processor, following programs, asks people what jobs are to be run. It is up to people to be sure that the proper programs and data are presented. In large computer facilities, this work is handled by the computer operations staff.

## Output

*Outputs* are the methods and/or documents that deliver needed information to users of computer systems. To be of value, output must have assured accuracy and be produced on schedule. Modern computer systems can deliver output in a number of forms, including:

- *Printed outputs* are documents prepared on machines that imprint letters, numbers, and symbols on paper. A wide range of printing devices is available for use with computers. These can be simple, typewriter-like units that imprint one character at a time. These can also be devices that

**Computer outputs** can take a number of forms. The photo above shows both a graphic display and printed graphic output on a plotter. At left is a high-speed laser printer that generates up to 20,000 lines of printed text per minute. (Photo at left, © Liane Enkelis)

imprint complete lines of data in a single operation. Some printers have mechanisms that strike the printing surface. Others form images electronically. *Impact printers* create impressions by striking a ribbon and paper. Electronic, nonimpact printing can be done by spraying ink onto paper. Another nonimpact method involves use of special paper that creates imprints from heat. Some high-speed machines print documents through use of drums like those used in high-speed copiers. Output speeds for printing devices can range from a few characters per second to as many as 20,000 lines per minute.

- Documents can be drawn on paper to provide *graphic output*. Graphic output can deliver any type of drawing or image, including letters, numbers, and symbols.

- Output can also be presented to users in the form of *displays*. A display is a temporary presentation of graphic or text information. The most common type of display used within computer systems is the screen of a *video display terminal (VDT)*. A video display presents information on a screen similar to that of a television set. The word terminal describes any device that is linked directly into a computer system and can be operated by users. Most terminals can be used for either input or output. A VDT has a keyboard for input. Entries into the keyboard are displayed on the screen. The computer can also display graphic or text data as outputs. Video display terminals are the most common form of direct output from computers to users. An advantage of video terminals for outputs is that users get their information directly and immediately from the computer. A disadvantage is that there is no printed record of the output. Therefore, some terminals include devices that can print, as well as display, their outputs.

- Computers can, in effect, speak to users through *audio output*. Devices are available that can output information through sounds that resemble closely the human voice. With audio output capability, any push-button telephone can become, in effect, a computer terminal. Users dial specific numbers to connect them to the computer. Then, they enter input codes that cause the computer to retrieve information. This information is then spoken to the user. An advantage of this approach is that many people can gain access to computer information. A disadvantage is that it is far slower to present data in spoken form than through other methods.

- Output text or graphics can be recorded on *microfilm* or magnetic media. Microfilm consists of a roll of a sheet of film on which special computer output devices record miniature images of text or documents. It is possible, for example, to place as many as 100 full-page images on a single sheet of film measuring four by six inches. Microfilm output is used in situations involving large volumes of output. Also, microfilm represents a permanent method of storing data. Magnetic media are the same as those described above. This output technique is used whenever information produced by a computer must be stored for later use.

**Microfilm saves space,** as demonstrated in this photo. The microfiche in the foreground stores the same number of pages of information as the stack of paper in the background.

As described above, computer outputs can be delivered by either *on-line* or *off-line* methods. On-line means, simply, that output devices are operated directly by users. Thus, users can inquire directly into computer-maintained files. Users can receive information from computers whenever they require these outputs.

*Batch outputs* are prepared in a computer center and delivered to users at a later time. Most batch outputs are printed or drawn documents. However, outputs recorded on microfilm are also handled in this way.

People are closely involved in computer output procedures. If on-line outputs are provided, users are in direct control. For batch service, personnel at the computer center are responsible for preparing the needed documents and delivering them. Responsibilities also include assurance that outputs are delivered on time and only to people who are supposed to receive them.

**Printed computer outputs** are valuable because they provide paper documents that can be reviewed and acted upon by people.

**Secondary storage media** include diskettes for microcomputers like the one shown here. In addition, disk packs and magnetic tape are also used on larger computers.

## Storage

Storage is sometimes known as *secondary storage.* The term secondary storage identifies devices such as tape or disk drives. These drives provide support and backup for a computer's main memory. In this sense, main memory is sometimes referred to as *primary storage.*

A major difference between memory and storage is that storage media hold recorded data for long periods of time. Data in memory are dynamic. These data are held to support processing. There is not enough capacity in memory to hold all of the data a system needs. Further, on most computers, data in memory disappear when the systems are turned off. By contrast, data recorded on magnetic media such as tapes or disks will remain there permanently.

Every computer system needs some kind of secondary storage. Secondary storage devices, typically, hold many of the processing programs to be used. In addition, data to support many applications are entered into and/or held on secondary storage. Individual computer systems may have widely ranging secondary storage capabilities. A personal computer may have the capacity of only a single diskette or cassette. On the other hand, large computer systems may have hundreds of disk or tape drives.

Computers provide an excellent example of how technologies build demands among people. The first electronic computer to be used commercially was installed in 1952. The customer was a General Electric plant in Louisville, Kentucky. At that time, it had been predicted that the whole world would use, perhaps, 300 computers. Today, more than 2 million computers are being manufactured each year.

## Summary

An information processing system is more than simply a computer, or even a computer system. People, data, procedures, and equipment are required to do the complete job of processing information.

People perform many roles in computer processing systems. For one, they are the users of these systems. People design the systems, write the programs, set up and operate the computers, input data, and monitor the actual work. Without people, computers would not exist—or need to exist.

Data are the raw materials for information systems, like iron is a raw material for making steel. Careful selection of data is important to the success of any information processing system.

Procedures must be uniform if an information system is to be useful. Job instructions tell people how to do a specific job. Programs tell computers how to do a specific job.

Equipment in an information processing system may include a computer. A computer is a set of devices that work together to process data. A computer may be so small that it fits into a briefcase, or it may be large enough to occupy a complete building.

Within a computer information system, the functions of input, processing, output, and storage, are automated.

The method for input depends upon the type of data used. Usually, data from source documents are changed into a form that the computer can use. Keyboards, similar to typewriter keyboards, are the most common input device used. Data can be recorded on many types of data media: punched cards, magnetic tape, disk, diskette, cassette, or encoded documents.

Processing takes place in a computer processor. Each processor has two sections: the central processing unit and memory. The central processing unit operates the computer and performs all arithmetic and comparison functions. The central processing unit also has two parts: the control unit and the arithmetic/logic unit. Memory holds the data and programs that are being processed at a given moment.

Outputs are either methods or actual documents that deliver information to users. Printed outputs are documents. Graphic output can produce drawings, letters, numbers, or symbols. Output can also deliver information through displays, such as on a video display terminal.

Storage is sometimes called secondary storage (with memory referred to as primary storage). The main difference between memory and storage is that storage media can hold recorded data for a long time and memory does not. Usually, memory disappears when the computer system is turned off. Information, data, and programs can be stored.

## Technology Terms

computer
program
computer system
systems analyst
computer operation
computer input operator
uniformity
manual
hardware
software
system software
application software
programming language

compiler

machine language

binary language

source transaction

source document

data capture

transcription

media

data media

storage media

data entry

direct input

automatic teller machine (ATM)

keystroking

punched card

magnetic tape

track

vertical column

bit

binary digit

parity bit

magnetic disk

retrieve

serial

fixed disk

disk drive

removable disk

magnetic diskette

magnetic cassette

universal product code (UPC)

optical character recognition (OCR)

magnetic ink character recognition (MICR)

processor

central processing unit

control unit

arithmetic/logic unit (ALU)

main memory

output

printed output

impact printer

graphic output

display

video display terminal

audio output

on-line

off-line

batch output

secondary storage

primary storage

## Using Technology

1. John is a computer input operator. He is inputting records that have 250 characters (letters, numbers, or other symbols). In his department, the input operators usually input an average of 10,000 characters per hour. How many records will John probably input in an eight-hour day?

2. What are the differences between the main memory and secondary storage? Write your answer in a short composition.

3. The data input methods used most frequently are keyboard entry to tape and direct reading of specially encoded documents. In a brief composition, compare these methods. What are the advantages and disadvantages of each technique?

# 14

# COMPUTERS AND CHANGE

### Your Learning Job

After you have read this chapter and completed the work assignments, you should be able to:

- ☐ Illustrate how technologies produce both good and bad results.
- ☐ Explain why skills and the ability to adapt are important in a technological society.
- ☐ Describe important effects that computers have had on workers.
- ☐ Identify several trends in computer technology and their probable effects on people in the future.
- ☐ Explain how computers have changed office procedures.
- ☐ Define the term robotics.
- ☐ Define artificial intelligence.

## TECHNOLOGY AND NEW IDEAS

New technologies succeed because people need or want the results they can deliver. Success for a technology or for a new idea using an old technology lies in helping people. Once people realize that help is available, change results. People do things differently because they believe the new ways are better. Work may become easier because of technology. Or, because of technology, people may be able to produce more, and earn more.

Putting it another way, people use technology to cause change. Change, in turn, comes from acceptance of technology by people. The technology itself doesn't change people. Only people change the way people do things.

This principle has been true throughout history. People who became tired carrying heavy loads tamed animals. As loads got

**Sizes and costs of computer equipment** keep getting smaller. One of the main reasons has been the introduction of integrated circuit technology. This photo shows the super-cleanliness used in the manufacture of integrated circuits. (Courtesy Storage Technology Corporation © 1983)

**Problems can result** as an effect of technology. This is demonstrated by the automotive junkyards that can be found near many cities. (Courtesy National Archives Trust Fund)

heavier, people invented the wheel. Hunters invented spears and knives.

As people built cities and became urbanized, the types of technologies they developed changed. In modern times, technologies began to involve trade-offs. A trade-off, where technology is concerned, is a situation that can involve both good and bad results. Every major technology has its costs:

- The automobile made transportation convenient and fast. But the automobile also led to accidents that kill 50,000 people and injure 1.7 million people each year. The automobile also pollutes the air while making it easier to move along the ground.
- Railroads made it easier to travel across the country. For most people, the ability to travel or to ship goods was a benefit. But, for those living near railroad tracks, trains meant dirt and noise.
- Electricity brought light and energy to homes, offices, and factories. Electricity has also become a major cause of accident and injury in homes, offices, and factories. In addition, many fires that take lives or destroy property have electrical origins.

The same type of situation surrounds any new technology. That is, technologies are accepted because they produce benefits. At the same time, technologies also result in costs and problems. People in government, therefore, must look at both the benefits and costs connected with technologies. If the costs or problems become greater than the benefits, controls are necessary.

An example of this kind of thinking can be seen in controls over automobiles. In 1973, the world was threatened with a shortage of oil. Until that time, the design of automobiles had moved steadily toward bigger, more powerful models. It was common to find automobiles that would travel only eight or nine miles on a gallon of gasoline. Government regulations were applied to save a critical resource—oil. Today, automobiles in the United States commonly travel from 25 to 50 miles on a gallon of gasoline.

Technologies can also have costs and benefits that cannot be controlled in this way. People must deal with such technologies by adjusting. These adjustments come through

**Improved output** in the production and distribution of goods can result from the use of computers. (Photo, courtesy of Hewlett-Packard Company)

knowledge, experience, and changes in the way things are done. One example is the relationship between technologies and jobs. Many technologies or inventions are created to improve *productivity*. Productivity is a measure of the amount of work that can be delivered by people or machines.

In general, the more work that can be completed in a given time, the less the finished goods will cost. Therefore, many technological inventions have been aimed at increasing productivity. The costs of the goods produced have come down. At the same time, however, many of the people who used to work at producing those goods have lost their jobs.

**When technologies change** people can be affected. With the introduction of the automobile, for example, people who used to care for horses and make harnesses and carriages had to be retrained to other occupations.

## Displacement of People

This factor of lost jobs has been—and will continue to be—a major trade-off with technology. As technologies are accepted, some people are *displaced.* Displaced means that people are put out of their jobs or homes. For centuries, displacement of workers has been a continuing effect of technology. Before automobiles, for example, most of the local transportation in large cities depended upon horses. Whole factories existed to make horseshoes, harnesses, and other devices needed in working with horses.

Many of the products—and the skills required to make them—were no longer needed. These changes in skill needs resulted as people switched from horses to cars and trucks for transportation. Cities like New York or Chicago no longer needed large numbers of people to make harnesses for horses. These people had to change the way they earned a living. They had to adapt to the new technologies.

As technologies have increased their impact, the adjustments of people have become more difficult. During the nineteenth century, for example, it was relatively easy for stagecoach drivers who lost their jobs because of trains to find new work.

**Changing technologies** are highlighted in this scene. The photo shows Henry Ford wheeling out the first automobile made in one of his early shops, on Bagley Avenue, Detroit. (The Bettmann Archive, Inc.)

They could go to work for the railroads. Or they could find jobs on farms or ranches.

Many later inventions, however, involved increasingly large numbers of people. People who were displaced by some modern inventions had a special problem. These people had been able to earn a livelihood even though they were *unskilled* or had only low skill levels. An unskilled person is one who has no special training for a specific type of job.

The impact of displacement upon unskilled workers has been particularly great in recent decades. The situation is illustrated easily. An important, necessary type of work in large cities has centered around sweeping the streets. Cities collect dirt. Street sweeping has been a necessary task to avoid both dirt and disease. Historically, street sweeping was an unskilled job. People with brooms swept the streets, each working in a small area. The dirt was gathered by hand and taken to collection points. There, the dirt was shoveled into trucks and carried away to be dumped.

During the 1950s and 1960s, many companies introduced street sweeping machines. These are devices mounted on trucks. Street sweepers are driven through the streets, picking up dirt

**Technology is fighting disease** in this photo. During an influenza epidemic in 1918, street sweepers in New York City wore face masks. During this era, cars and trucks were replacing horses, eliminating germ-carrying manure. (The Bettmann Archive, Inc.)

rapidly and in large quantities. A single street sweeping machine cleans large areas of a city. The area covered by one street sweeping machine might have required the services of hundreds of people with brooms. These people with brooms became displaced. Jobs simply ceased to exist.

Of course, other jobs were created. People were needed to make the street sweepers. Other people were needed to maintain (service) the machines. An entire replacement parts market developed. That is, people were needed to make the tires, brushes, and other street sweeper parts. Of course, still other people were needed to drive the street sweeping machines.

However, almost all of the newly created jobs involved skills. By contrast, the people who used to push brooms could be unskilled. As developments like the street sweeper were introduced, fewer and fewer unskilled jobs were left.

In general, the people who have been hardest hit, or displaced, by technology have been the unskilled workers. People without skills are relatively helpless as technology continues to develop. Technologies that are accepted usually benefit many people. The benefits have tended to outweigh many of the costs of displacement. For example, a city can save millions of dollars in operating costs by using street sweeping machines. Most of the people who live in the city—and pay taxes—save money through the use of these machines. Therefore, it is considered that the benefits of these savings are greater than the costs of worker displacement.

## Value of Transferable Skills

At the same time, technologies can be a major benefit to workers who have *transferable skills.* A person working as an auto mechanic, for example, might get a better job working on street sweepers. A street sweeper is more expensive than an automobile. Therefore, a person who can learn to take care of a street sweeper often can earn more. For such a person, a street sweeper is an opportunity. However, to take advantage of the opportunity, the person needs some transferable skills. A mechanic who knows how to fix engines, for example, can move easily from work on trucks to street sweepers.

**Computer-related skills** can be valuable background for persons entering the job market. Many schools are teaching the use of computers to students.

This type of situation represents an important reason why you are studying technology today. People who work must continually adapt to new technologies to qualify for different types of jobs. The idea of studying technology, then, is to understand the principles that are applied in developing new methods. With such an understanding of principles, it becomes possible to deal with—even to benefit from—change.

People who build a foundation of knowledge and a basis of skills become *adaptable.* Adaptable means, simply, that you are able to understand the reasons for change and to adjust your thinking and behavior. For an adaptable person, changes usually represent opportunities. For people who are *inflexible,* or not adaptable, change can be a threat. Since there is little that individual people can do to stop change, inflexible people tend to be unhappy.

Even in the face of major changes, people who understand the basics of technology and who have developed basic skills find opportunities. These values of knowledge and skill are easily illustrated. Consider what happened, for example, when

**A major impact** upon the lives of many people was created with the introduction of jet aircraft. Within a very short time, jets replaced propeller-driven airplanes. This made it necessary for aviation workers to master entirely new sets of skills. (American Airlines photo)

jet airliners were introduced to commercial aviation. The first jet airliners went into service in 1959. Within a few years, older, propeller-driven aircraft virtually disappeared from commercial passenger service.

In a short span of time, many thousands of skilled people adapted to new opportunities. Experienced pilots, for example, had a broad knowledge of the principles of flying. They were able to adapt easily to jets. If a person knew the basics, a jet engine was simply a different source of power. With comparatively little training, pilots found themselves earning more money. The pilots were worth more because they met the challenges of the new, larger aircraft.

Making the advanced aircraft created many job opportunities for drafters, engineers, and people with metalworking skills. Change resulted in thousands of opportunities for workers. At the same time, jet aircraft were rapidly accepted by the public. Passengers could fly farther, in a shorter time, for less money. The public was served. Skilled people adapted and improved their situations.

As another example, consider a change that has taken place in offices in recent decades. For many, many years, shorthand had been considered a basic office skill. Shorthand is a method of using symbols to represent words. A person who takes shorthand is called a stenographer. Stenographers were able to write shorthand notes as rapidly as people talked. Large volumes of business correspondence were handled through shorthand. An executive spoke, or dictated, the text of a letter or other document. A stenographer wrote the words in shorthand. Then the stenographer typed the document on a typewriter. This involved reading the notes and typing documents that contained the spoken words.

From the 1950s and 1960s onward, a new technology, dictation machines, has displaced much stenography. With pocket-sized machines, executives can dictate letters or other documents at any time. With stenography, it is necessary for two people to be present. One person speaks; the other makes notes. With machines, executives have found that they can do dictation at odd hours. Some executives dictate on airplanes or in cars. Shorthand would not be convenient in these situations. Thus, the executives can get more done. Transcription takes place from the recordings, rather than from stenographic notes.

Because stenographers developed important, basic skills, dictation machines have caused virtually no displacement. Rather, the value of stenographic training has been transferred to other areas of business education. Experience showed that the writing of shorthand is simply an alternate method of recording dictation. The real skills involved were in being able to listen to and understand a broad business vocabulary. Further, a good stenographer had to be a good speller. A good stenographer also needed excellent language skills.

There was more to transcribing business documents, experience showed, than simply operating dictation machines. Today, business education programs are placing increasing emphasis on language skills and spelling. In turn, people who have these basic skills are assured that there will be interesting, rewarding jobs open to them.

## THE EFFECTS OF COMPUTERS

Similar situations have occurred in information technology. People who have built a sound basis of knowledge and skills in using computers have found security and opportunity. Notice the emphasis on the idea of using computers. Computer technology is one of the fastest changing areas in all of human history.

People who work on the design or building of computers must continually re-educate themselves. This is a challenge in

**In modern offices,** the need for shorthand skills is diminishing as dictation and transcription machines come into use. (Courtesy Dictaphone Corporation)

**Type composition** is an example of a technology that was slow to change. The same basic methods developed by Gutenberg in 1452 remained in use for more than 400 years.

itself. For example, the *half-life* of an electronic engineer or technician is perhaps four or five years. Half-life has a special meaning in relation to knowledge. Half-life is the time it takes for half of everything a person knows to become obsolete. Obsolete means that an item of knowledge has been displaced. Obsolete facts are replaced by new ones.

In electronics and in computers, half-lives are extremely short. A half-life of four or five years means that a person must study continually to keep up with new developments. A person who did not study continuously would not be able to hold the same job after four or five years.

On the other hand, there are many skills and knowledge areas associated with computers that are not threatened by short half-lives. These longer-life skills involve the adapting of computers to meet the needs of people. There are many specialists who study the needs of computer users. Their job is to understand what users want, then design information processing answers to those needs. In general, persons who understand both the needs of people and the capabilities of computers are

highly adaptable. By comparison, people who specialize in technologies alone must work continually to keep up. Such people can find it harder to adapt to change.

It is stated above that computers have been the most rapidly changing technology in human history. Earlier in this book, examples of long-lived technologies have been cited. For instance, Gutenberg invented moveable type in 1452. The same basic technology was used in setting type for more than five centuries. Keyboard operated typesetters were introduced in the 1880s. These machines dominated the market for some seventy years. Then, when computers were introduced into typesetting, everything was revolutionized in less than a decade.

The history of the computer field has been broken down into a series of "generations." In computer terms, there has been a new generation of equipment every seven years since the early 1950s. A generation, for computer specialists, occurs each time there is a major technological breakthrough. Breakthroughs of this type have been measured in terms of *orders of magnitude.* An order of magnitude means that a new method is, by some

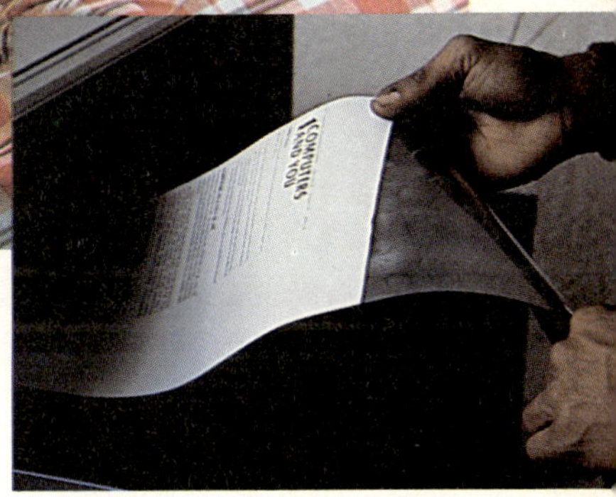

**Computerized typesetting** was accepted rapidly following introduction of this technology. Under modern techniques, text to be typeset is captured in computers, as shown above. The computer then processes finished type automatically, as shown at right.

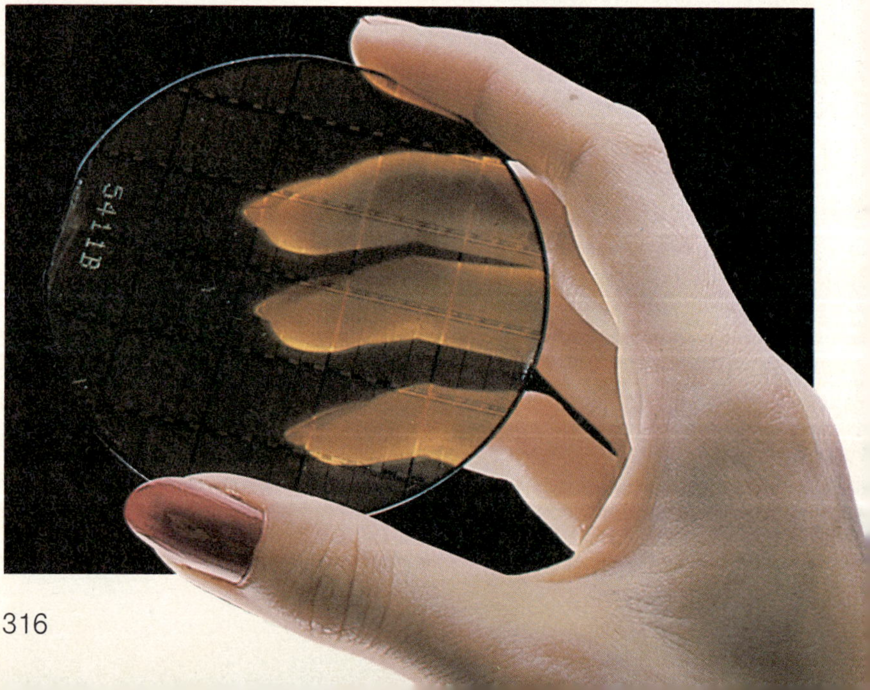


**Computer memory devices** offer an example of rapid change in technology. As late as the early 1960s, computers had large magnetic cores for memories, as shown in the photo above. Today, thousands of times as much capacity can be contained in small wafers like the one shown at right. The newer memories also operate faster because they are smaller.
(Photo at right, courtesy Intel Corporation)

measure, at least 1,000 times better than the old one. For computer processors, each new generation has been 1,000 times faster than the old. The same has been true for memory access operations. Also, with each new generation, costs of processing or data storage have been reduced at least tenfold. That is, with each new generation of secondary storage devices, unit costs are one-tenth of what they were. For early disk devices, for example, storage costs averaged 10 cents per bit of data. For the next generation of equipment, costs were reduced to 1 cent. Then, they came down again to 0.1 cent per bit.

With technology advancing at this rate, computer power has become increasingly available—at far lower costs. Lower costs, in turn, have made it possible to use computers for increasing numbers of applications (jobs). Computer specialists have been at work continually to develop new areas for computer use. Thanks to these efforts, a number of areas of future, or growing, computer uses can be identified. These future uses of computers can be important to you. This is because many of these new computer applications will develop during your working career. You will live with the changes resulting from these new computer applications. Some of the more important of these new computer applications include:

- Personal services
- Office automation
- Robotics
- Ergonomics
- Artificial intelligence.

**Microchips** used on modern computers are small in size, large in capacity. Devices of this type have helped to increase performance while reducing prices for computer equipment. (Courtesy IBM Corporation)

## PERSONAL SERVICES

Microchips have made it possible to build computer capabilities into almost any type of equipment. Thus, computers can be serving you regularly without your even being aware of them. A microchip, as you know, is a tiny electronic device. Complete computer processors can be built upon microchips. This means that you are already using computer processors that are smaller than one of your fingernails. Computer memories can also be reduced to this size.

As examples, ovens and microwave cookers with automatic controls use microcomputers. These microcomputers make it possible to enter instructions that cause food to be cooked automatically. Modern dishwashers, washing machines, and clothes dryers also use microprocessors.

**Automatic manufacturing** capabilities have helped increase supplies and reduce costs of computer microchips. (Courtesy Storage Technology Corporation © 1983)

Computers are also a major part of entertainment equipment in many homes. To illustrate, a video recording device puts a computer to work for you. You can enter instructions into the video recorder about what channels you want to record, at what times. The computer in the recording unit carries out your wishes.

The family automobile is also becoming a highly computerized piece of machinery. Separate computing devices monitor fuel consumption, engine functions, brakes, and many other operations on modern cars. It is even becoming possible for the computers in your car to communicate with larger computers in garages. Cars of the future will have computer plug-in points. These will be used to connect computers in garages that check the operation of your car. Without ever lifting the hood, you will be able to start the engine and have the computer analyze your car. These special computers check all working systems within the car. They report on any problems or potential problems. That is, even if a problem does not yet exist, the computer will be able to spot it in advance.

If you have a personal computer in your home, it can work for you while the family is gone. Personal computers can be used

**Personal services,** such as the ability to shop right from home, are becoming available. The system shown in this photo uses a combination of a televison set and a computerized control network.

as the basis for home protection and energy systems. A home protection system sets off an alarm if an unauthorized person enters your home while you are away. The same computer can also monitor temperatures in your home, turning heating or cooling equipment on or off.

Your personal computer can also become a personal shopping service in the near future. Banks, stores, and other businesses will offer these services. You will be able to call special numbers that link your computer to computers in business organizations. With these computers, you will be able to order merchandise. You will also be able to arrange for a bank to pay bills from your account.

Your personal computer will also admit you to many important sources of knowledge. Services already exist that make it possible for people with personal computers to use special libraries or *data banks.* A data bank is, simply, an information storehouse that can be used with computers. With these information services, any computer user can look at vast amounts of text materials. For example, entire encyclopedias can be referenced by computer.

It is interesting to note that none of these home or personal applications represent needs of people. Rather, all of these applications deal with wants—with conveniences for people. As people are willing to pay more for conveniences, computers will provide more services. Tens of millions of computer processors will be built and installed for these purposes. In turn, thousands of job opportunities will result.

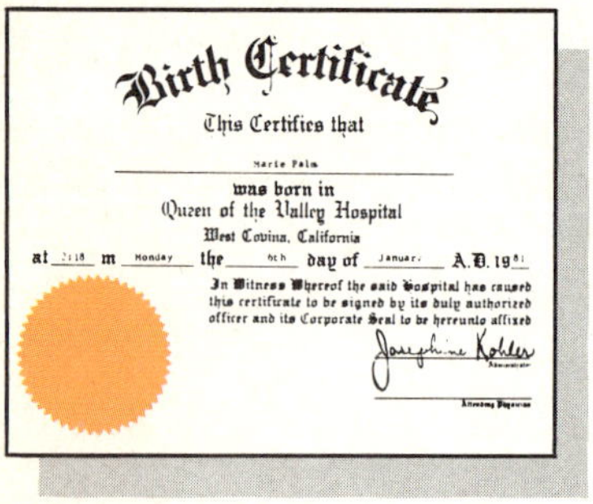

Birth Certificate

This Certifies that

was born in

Queen of the Valley Hospital

West Covina, California

at ___ m ___ the ___ day of ___ A.D. 19__

In Witness Whereof the said Hospital has caused this certificate to be signed by its duly authorized officer and its Corporate Seal to be hereunto affixed

**Population statistics,** including the production of birth certificates, are jobs suited ideally for computers.

## OFFICE AUTOMATION

Offices of large business and governmental organizations are paperwork centers. Pieces of paper that are handled represent millions of different transactions. Pieces of paper record almost everything in the life of every citizen. A piece of paper is created when you are born. Other paperwork results when you receive shots to protect you against disease. In school, your activities create a trail of paperwork.

Increasingly, pieces of paper are being replaced by video screens. Large collections of paper are being replaced by computer storage devices. Offices are becoming electronic. In the process, the work being done in offices, and the kinds of people who work in offices, are changing.

Under traditional methods, office workers had to do—then redo—the same jobs many times. To illustrate, consider such a simple task as writing a letter. An executive dictates a letter. A transcriptionist types the letter from the executive's cassette. In the great majority of cases, the letter will have to be changed. Perhaps the transcriptionist typed a word wrong. Just as often, the executive will want to make changes after reading the document.

With conventional typing methods, this meant that the entire letter had to be redone. With computerized word or text processing programs, only the changes have to be reworked. The texts of documents are kept on secondary storage media. These texts can be recalled, changed, then stored in the new form at any time. The texts can also be printed on computer output devices as many times as needed. Thus, where a typical letter might have required three, four, or even six typings, this rework is now unnecessary. Computers can generate new documents as fast as people tell them to do so.

In a modern office, it is possible for many computers to talk to each other. Special connections, called local networks, are linking computers with one another. The computers involved may be small or large. For example, in some companies, hundreds of small, personal computers used

**Office practices** used to require considerable repetition of work tasks. Computers have changed work styles while styles in clothing and furniture were also changing.

by executives are linked to large central computer systems. This makes it possible for authorized executives to look at information representing the company as a whole. Authorized executives can even add to or change computer files.

**Business executives** are, increasingly, keyboarding their own messages into computer systems. This practice is changing the skill requirements for workers in modern offices.

People in business are changing the way they look at paperwork jobs. For example, an important trend in business today is *executive keyboarding.* This means, simply, that many executives have computer terminals in their own offices. Instead of having to call someone to get information they need, executives can retrieve records themselves. Executives can also write short memos or letters right at their own keyboards.

The value of such computer services is not restricted to individual offices. A growing area of business services is *electronic mail.* This means, simply, that computers can send information to each other over long distances, electronically. It is proving both faster and less expensive to use electronic transmission in many cases than to mail letters. Normally, a letter has to be typed, rechecked, placed in an envelope, and mailed. By contrast, the content of electronic mail can be checked right on an executive's terminal or computer. Then, the message can be sent to a single destination or to thousands of destinations. All this can be done in a fraction of a second.

People who have electronic mailboxes check from time to time to see if they have any messages. These messages can then be called up and read on computer screens. Anyone, at any time, can cause a computer to print a copy of any electronic mail document. However, only needed documents are printed. If only the information is needed, the cost of printing and mailing can be saved.

Many business jobs require large, formal documents. For example, consider what happens when a company buys a computer. The computer will come with thousands of pages of manuals. There will be separate sets of instructions for the people who maintain hardware. Programmers will have other instructions. Still other instructions will be provided for users, and so on.

**Typeset documents** can now be created directly on electronic printers connected to computers. This photo was taken at a demonstration of a newly introduced electronic printing system.

Computer companies can now use computers to create these documents. Texts are entered into secondary storage devices. Special output units can now produce these in printed formats, as though they were typeset. This is done with electronic printers. These electronic printers can produce graphic output as well as material that looks as though it came out of a book. This entire field is being referred to as *electronic publishing*. Instead of having to print hundreds, or thousands, of sets of instructions, companies can print one or a few at a time. The texts and images are stored in computers. They are then printed out at high speeds—only as they are needed. Major savings can be realized in the cost of preparing this type of document.

In modern offices, it has become possible to hold meetings and conferences electronically. The people taking part in the meeting may be thousands of miles apart. They are brought together by video or computer systems. People can actually show each other documents. They can see one another and hold discussions as though they were in the same room. The people involved save travel time. Their organizations save the increased costs of having people travel to and stay in distant cities.

**Working at home** is a new option being opened by computers. Instead of traveling to offices, people can work at their own computers, in their own homes. These computers are connected to computers in company offices. Thus, people can interact with others in their organization without leaving home.

One possible trend in office procedures is for people to stay away from their offices. That is, with computers, people can work right in their own homes. Their personal computers can be linked with computers in the office. Work can be assigned and delivered through computer networks.

During your working career, you will undoubtedly see many changes in the way people work and live. Many of these changes will represent opportunities. The majority of the changes will probably be computer-related.

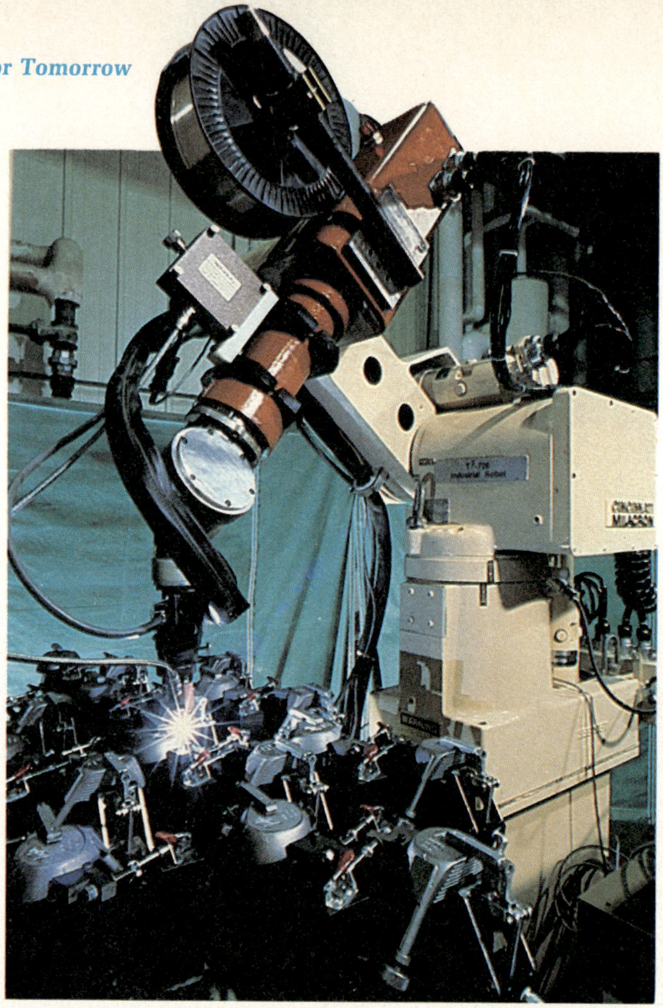

**Computers assist in design and production** within modern organizations. Photo below shows use of a computer for product design. At right are robotic devices operated by computers. (Photo below, courtesy of Hewlett-Packard Company. Photo at right, Courtesy Cincinnati Milacron.)

## ROBOTICS

The term *robotics* stems from the word *robot.* A robot is any machine that operates by itself and performs some task. Robotics involves the use of equipment controlled by computers to perform a series of jobs. Robotics equipment handles major jobs, usually with great accuracy.

Projects can range from individual machines that run themselves to complete factories. For example, entire car bodies are now being made automatically. Robotics equipment brings all of the parts together. Then robotic welders join the parts. As many as 300 welds can be applied in a matter of minutes. After these processes are completed, the robotics units inspect their own work.

Robotics equipment is also used for the stocking and delivery of equipment parts. Computer entries tell the equipment

**Metal manufacturing systems** can be operated automatically by computer-controlled machines.
(Courtesy McDonnell Douglas, St. Louis)

where to place needed parts for products to be manufactured. Other computer entries select the parts when they are needed and deliver them to the points where they will be used.

The use of robotics is related closely to two other advanced computer techniques. These are *computer-assisted design (CAD)* and *computer-assisted manufacturing (CAM)*. In effect, robotics represents an extension of CAD/CAM techniques. Engineers using CAD programs can actually design devices or machines through the use of computers. The computer translates the lines entered by designers into numeric values, using numbers to draw pictures.

Once a design is developed under CAD techniques, CAM programs can be used to drive the machines to make the needed parts. Manufacturing devices driven by computers are used to shape and process metal parts. CAM devices also make precise electronic components. In effect, robotics represents large-scale use of CAD/CAM techniques.

## ERGONOMICS

*Ergonomics* is a technological specialty also known as *human engineering.* The field involves the design of working equipment for the most comfortable and efficient use by people. The field of ergonomics can be divided into two broad categories.

- Work situations
- Human performance.

### Work Situations

Ergonomics specialists study people at work. Their goal is to make work easier, more comfortable, and more productive. Many ergonomics studies have dealt with the use of computers. For example, the use of computer terminals in offices represented a change in work situations. Working at a computer terminal is different than working at a typewriter. For one thing, a terminal operator spends a lot of time looking at a video screen.

The height at which the screen is placed, or the angle at which it is viewed, can affect a person's comfort. Some people

**The importance of ergonomics** is highlighted with an extreme example at this display showing a skeleton at a computer work station. The company using this display produces computer work stations and chairs.

**A modern office work station** includes a computer terminal with a conveniently located keyboard. Other requirements for the worker are within easy reach. The entire area is pleasantly lighted.

feel that reading illuminated (lighted) displays on screens is harder on the eyes than looking at paper documents in typewriters. Other people have felt that they get stiff necks from looking up, or down, at computer screens. The types of chairs used by computer operators have also been studied.

Studies of this type have led to many changes and recommendations about working with computers. For example, because computer screens are lighted, many people used to work in darkened rooms. Under these conditions, it was found, the brightness of the screen can cause eye strain. Many computer terminals now come with movable screens. With these, the operator can change the angle of the view for comfort. Also, many computer terminals now have keyboards that can be separated from the screens. Again, the idea is to provide flexibility to make the operator comfortable. Many computer work

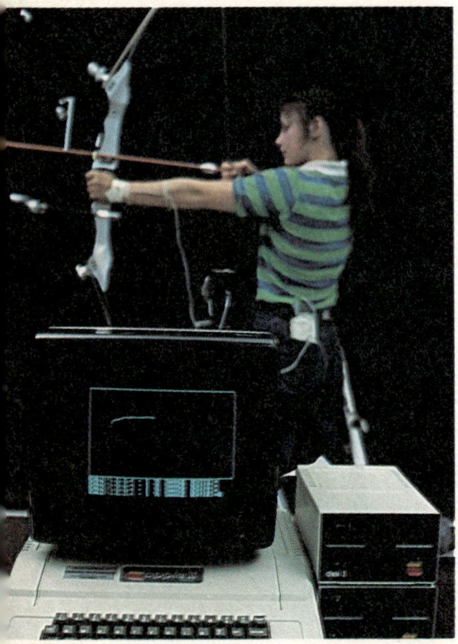

**Athletic performance** can be analyzed by a computer used as a training aid. (© Brian Payne/West Stock)

stations make it possible to adjust the height of desks or tables. The principle is simple: People come in different sizes and heights. So, the facilities at which they work should be adjustable to the needs of their bodies.

### Human Performance

Another dimension of ergonomics lies in studying the human body at work. Possibly the most dramatic use of computers for this purpose has occurred in the training of world-class athletes. People training for the Olympics or other major athletic events are studied with the aid of computers. Photographs of these athletes in action are translated to images similar to those used in CAD programs. In effect, motion pictures are translated to lines representing muscles, bones, and motion. These studies can then be shown to the athletes themselves. The athletes can then adjust their performance patterns for better results.

## ARTIFICIAL INTELLIGENCE

*Artificial intelligence* is a term describing computer systems that alter their processing on the basis of experience. In effect, computers are programmed to study their own data and information. Based on these studies, computers are programmed to change their reactions to situations.

As an example, consider a computer-operated inventory management system. Systems of this type are set up initially with program controls. The programmers are told what usage is expected for each part. Then, programs include allowances for the number of items to be used on a weekly, monthly, or other basis. Programs also develop estimates on how long it will take to order inventory items and receive new supplies.

Usually, inventory systems produce outputs that are reviewed by people. People decide when to place orders. These decisions are based on evaluations of actual use of items and experience in buying and receiving new supplies. With advanced, artificial intelligence programs, the computer can be directed to build and use its own information. Computers can establish minimum time requirements for ordering supplies and receiving replacements. Computers can monitor use of inventory items on a day-to-day basis. Computers can then "decide" when to order new products and how many items to order.

**Massive computer memories** can be constructed by interconnecting large numbers of microchips. (Courtesy Storage Technology Corporation © 1983)

As another example of artificial intelligence, computers are now being developed to recognize characters and numbers in printed documents. A modern computer system, for example, can scan pages of printed books. This scanning can result in a reading of the documents and recording of texts in memory. The computer that does this has a general program that helps recognize letters, numbers, or special symbols. The computer displays a picture of any letter, number, or symbol that it does not recognize. A human operator then tells the computer what letter, number, or symbol is represented by that image. The computer remembers this image. Thus, a computer can be "taught" to read any printed alphabet, or type font. Once a computer has been told what a certain symbol means, it will "remember" that meaning.

Artificial intelligence, then, means that computers can be programmed to learn, and to react to what they learn. The programs, however, are still developed by people. Thus, in artificial intelligence, as in other computer applications, the key lies in the ability of people. People have to originate and implement ideas. Technologies are developed by people, for people. This principle will not change. However, almost everything people do will change at some time in the future because of the results of technologies.

## Summary

The costs and benefits of technology must be considered by people. One troublesome result of many technologies has been the displacement of workers. Many people who have few job skills are finding that their jobs have been taken over by machines.

In a technological society, people need skills to get and hold jobs. Also, people need to understand the principles of technology to prepare themselves for changes. Being adaptable and learning to develop skills will help people to fit into a technological society. Computer technology is changing the way people work and live.

Computer technology already serves you in many ways you may not realize. The microchip is now used in many household appliances, entertainment devices, and automobiles. Personal computers can be programmed as a home alarm system. People will also be able to shop by computer.

In offices, computers are changing the way things are done. Paper records are being replaced by video screens. Communication between computers is possible now. With computers, people will be able to stay at home and still work.

Robotic equipment uses electronics to perform tasks. Robotics can be used in assembly line work, such as automobile manufacturing. Robotics represent large-scale use of CAD/CAM techniques.

Ergonomic technology is involved with the comfort and work performance of the human body. By understanding the problems that people who work with machines have, ergonomic specialists can suggest changes.

Some computers are programmed to learn. People, of course, still control the data input that the computer uses to learn. However, computers can be programmed to react to what they learn. This field is called artificial intelligence.

## Using Technology

1. How might computer technology affect schools in the future? Write a description of possible changes in school systems. Base your description on current trends in computer technology.

2. What existing jobs today might not be available in the future? Why is it possible that these particular jobs might disappear? Write your answer in a short composition. Again, get help from magazines you have at home or can find in a library.

3. What are the future trends in the job market? Find out from library sources what kinds of work will need large numbers of people in the future. Share your findings in class.

## Technology Terms

productivity
displace
unskilled
transferable skill
adaptable
inflexible
half-life
orders of magnitude
data bank
executive keyboarding
electronic mail
electronic publishing
robotics
robot
computer-assisted design (CAD)
computer-assisted manufacturing (CAM)
ergonomics
human engineering
artificial intelligence

# PUTTING TECHNOLOGY TO WORK

Technologies are created by people to meet the needs or wants of people. Creating technologies is a complex process. First, the needs or wants to be met have to be defined. These needs or wants must be understood clearly before development of a new product of technology can begin. Then, considerable amounts of information and knowledge must be focused on the situation. This is a special requirement for technologies. Science can result from discovery. But technology requires careful development.

Chapter 15 discusses the development of technology. The chapter outlines a project structure—a set of standard procedures. These standard steps, followed in order, can guide people in the development of technological products or results. In turn, understanding how technologies are created can contribute to your success in a technological world.

Chapter 16 reviews, broadly, the effects of technologies on people and their life-styles. The effects of technology upon a number of job areas and upon personal lives of people are discussed.

# 15

# THE SYSTEMS APPROACH: PUTTING TECHNOLOGY TO WORK

### Your Learning Job

After you have read this chapter and completed the work assignments, you should be able to:

☐ Describe the differences between pure and applied science.

☐ Define a system of technology through examples.

☐ Identify the parts of a production-type system and explain the relationships among those parts.

☐ Trace the steps you might take in applying a systems approach to a personal project.

☐ Ask a series of questions that would help evaluate the feasibility of a project.

☐ Outline the elements of a plan of work.

☐ Explain the role and importance of feedback in any systems approach.

☐ Contrast general design and technical design of systems and explain how one affects the other.

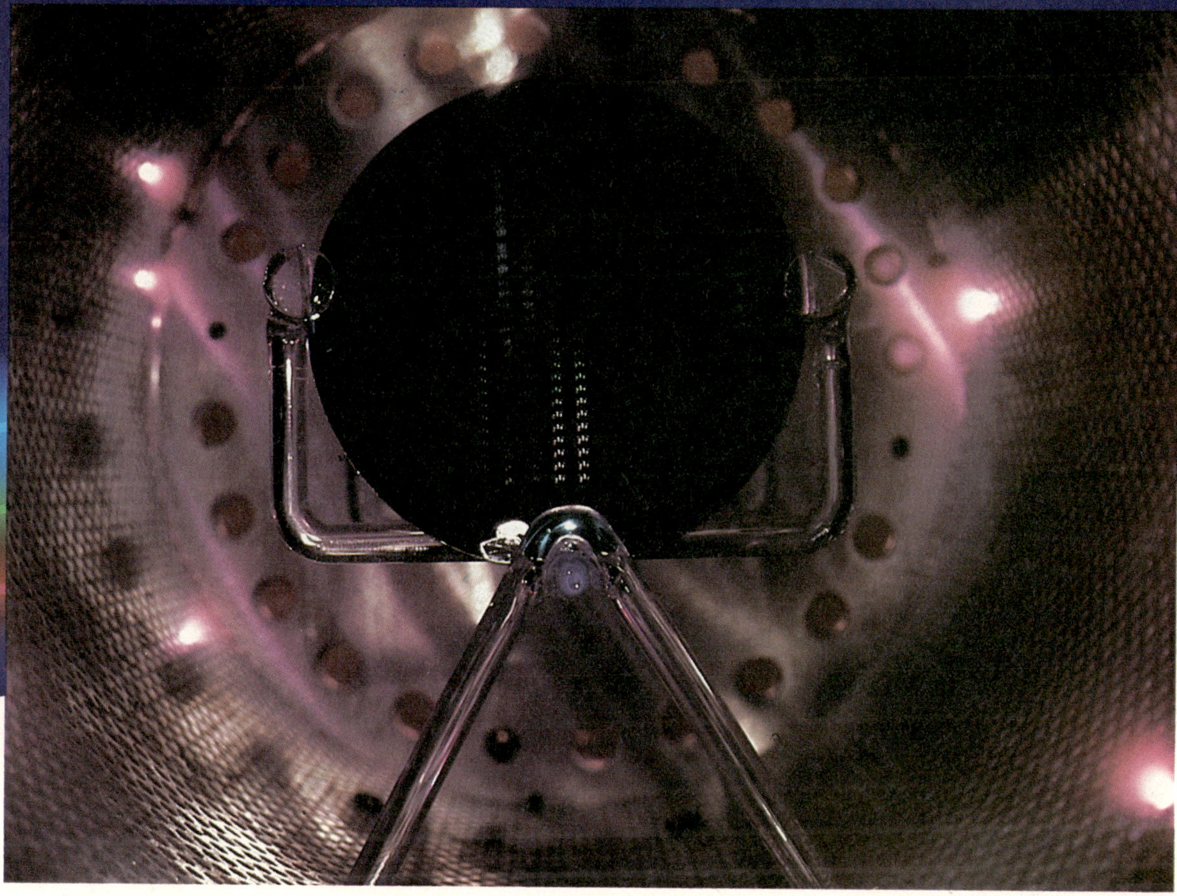

## DEVELOPMENT OF TECHNOLOGIES

**Technology puts the principles of science** to work for people. This manufacturing facility, that produces wafers for use in making microchips, provides an excellent example of this principle. (Courtesy Storage Technology Corporation)

Technologies put the principles of science to work for people. This means, in effect, that the principles of science come first. It has to be this way for technologies to apply those principles. It also follows that science and technology are related. However, science and technology have some major differences. These differences center around the purposes, or missions, of science and technology.

Science grows out of discoveries by or imaginations of people. Scientists observe, they state theories, then they set out to test those theories. A pure scientist is out to build knowledge through exploration. Science has no specific boundaries, goals, or targets.

There are differences, however, between *pure science* and *applied science.* People engaged in pure science do not have any specific end product in mind for their work. Their purpose is discovery. Applied science, on the other hand, is directed

**Analysis** is an area in which scientific research and computer technology are joined. The computer has become an important research tool. (Four By Five, Inc.)

toward using scientific principles for specific results. Applied scientists meet needs and wants of people by developing products.

Applied science results in technology. Efforts at developing technologies are specifically directed toward end results. New technologies result from the work of teams of people, equipment, procedures, and materials. Bringing these elements together to produce specific end results is known as the *systems approach.* That is, a system delivers known results from the efforts of people. These people follow procedures and use materials and equipment in specific ways.

The systems approach is also known as *systems development.* All technologies result from systems development methods. This chapter reviews the principles of the systems approach and of systems development.

## SYSTEMS OF TECHNOLOGY: TEAMWORK IN USING KNOWLEDGE

The machines and services that support modern life use combinations, or systems, of technologies. It takes these combinations, or systems, of technology to perform the complex tasks required by modern society. Consider just a few examples:

- There are more than 50 million cars and trucks on the roads in the United States. People have come to accept the private automobile and to depend upon it. Yet, each car or truck that you see represents a complex combination of technologies. The technologies of energy and electricity are present in every car. You need electricity to start the car and to keep it running. The technology of combustion is applied to burn fuel as a source of power. Structural technologies are applied to build the body of your car. Special technologies are at work in the cooling and lubrication systems. To steer your car, you apply a technology known as *hydraulics.* This technology specializes in deriving force from the control and movement of liquids. A technology known as *pneumatics* is put to work in most shock absorbers. Pneumatics deals with the behavior of gases or air under pressure. Pneumatic shock absorbers help smooth your car's ride. The tires on which the car rides are also pneumatic devices; they are air-filled. The materials used to provide glass for and to paint

your car also result from individual technologies. Before a new car ever goes into production, thousands of technical experts have contributed to its design. Then, production technologies take over. Thousands of additional people become involved.

**Systems of technology** are put to work to build products that people take for granted. For example, multiple systems of technology are in use in every passenger car. (Courtesy Ford Motor Company)

- To bring even a small computer to market, a large number of technical people must be involved. Some people work on the microchips that form the processors and memories of computers. Others work on secondary storage devices, such as disk files. Other, highly trained specialists worry about how the parts will go together. These people are often called *computer architects*. Even after all these efforts, remember that a computer itself is a collection of wires and components. Often, more people are involved in developing software than in designing and building hardware. Computer

**Many people with technical skills** are needed to bring even a small computer to market. (Courtesy Sperry Corporation)

software has grown so big that it involves multiple technologies on its own. There are specialists in systems software and in applications software. Each of these areas, in turn, has many subspecialties. Computer manufacturing also is a high-technology area.

- The airplane uses all of the technologies described in connection with both automobiles and computers. An airplane has structures, uses combustion, and needs shock absorbers and rubber tires. In addition, airplanes use computers as part of their navigation systems. Airplanes also have extensive communication systems. Much of the electronics industry today was born in the research that developed radar. Radar was created during World War II as a means of tracking aircraft. Still other technologies are involved in building and operating airports. The reservation systems used in selling airplane tickets represents advanced, large-scale computer applications. Modern aircraft take years to design, build, and test—even when computer-assisted design tools are used.

Reliance on technology becomes even greater if you think about and discuss space travel. Space travel does illustrate an important characteristic of technology. Think about the broadcasts you have heard or the articles you have read about space travel. A space shuttle, for example, lifts off from a complex base in Florida. The shuttle's flight is controlled out of a high-technology center in Houston. In flight, the shuttle communicates by relaying signals through communications ground stations all around the world. Ultimately, the shuttle lands in Florida or California after being tracked by a network of ships and planes. Many thousands of people are involved. Without close cooperation among all of these specialists, space travel would be impossible.

Teamwork—close coordination of efforts of specialized people—is essential to the development and operation of systems of technology. This is what the systems approach is all about.

## THE SYSTEMS APPROACH

Teamwork is vital because all of the parts of any system of technology must function together to produce needed results. Any system that produces results follows a series of steps that are carried out in order. Satisfactory completion of all the steps makes for a successful system.

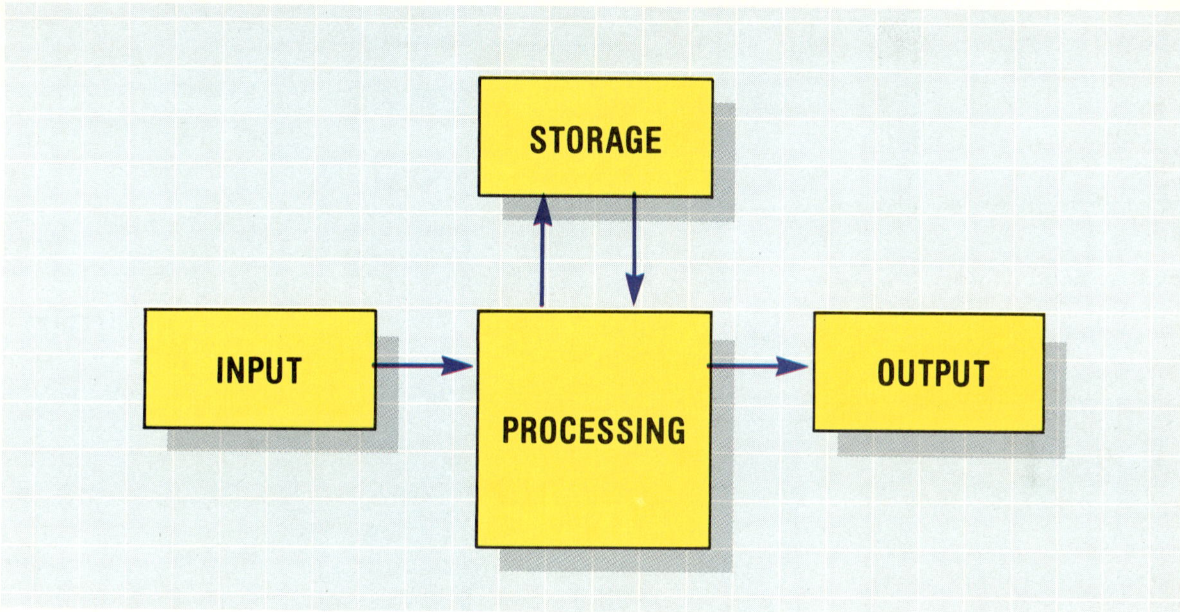


This approach, involving a series of specific steps, applies to all systems. Chapter 12 illustrates the steps in one type of system—an information system that processes data and delivers information. The functions of an information system, you will recall, include input, processing, output, and storage. The diagram of these steps is repeated from Chapter 12 to help you recall this system structure. A diagram of this type, that illustrates how a system operates, is sometimes called a *model.*

**Information system steps** include input, processing, output, and storage.

The same general approach applies for any system that involves use of technology. Only minor adjustments are needed in the basic diagram to show the functions of other types of systems. For example, look at the diagram, or model, for a manufacturing system on the next page. There are still input, processing, and output functions. A difference lies in the change of position and name for the storage function. In an information system, data and information storage relate to processing. In a manufacturing system, the stored items are called supplies. These are the basic materials used to produce a product. These materials feed into the input function, then to processing.

Another added element in the manufacturing model is the *monitor* function. The monitor function is set up to keep track of what is happening. Operations as complex as manufacturing systems can develop unexpected problems. When problems or

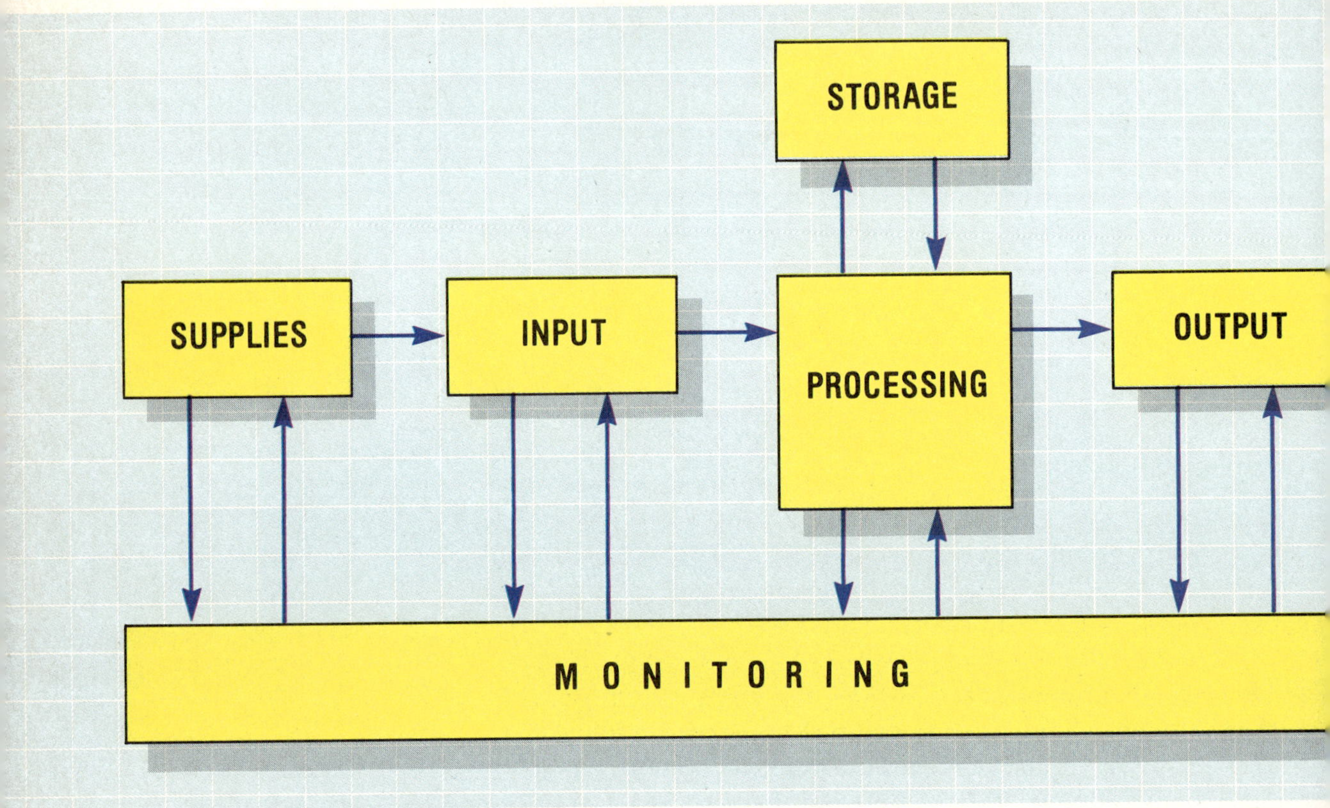


**The model for a manufacturing system** resembles that of an information system. Additional requirements include the stocking of supplies prior to input. Also, monitoring of progress is an essential part of all production systems.

special needs occur, they are reported to or detected by the monitor function. Corrections or adjustments are made as necessary.

Note that the monitor function is present in all systems that produce usable, reliable results. Information systems require monitor functions, even though Chapter 12 didn't include this detail.

To illustrate the idea of a monitor function, consider an athletic event such as a basketball game. Two teams of five players each use the resources they have available. Each team is after the same result, winning the game. Many of the actions and results during a basketball game are unexpected. Some can cause problems. Thus, a control or monitoring function is needed to keep the event running smoothly. Monitoring is done by referees and coaches. If players commit actions that are against the rules, the referees call fouls. As players become tired or are hurt, the coaches make substitutions. The coaches also use the human resources of their teams in the way they consider best. This means that the coaches are always making changes or adjustments in their lineups.

Now apply these same ideas to a business-type situation. Suppose you and some fellow students joined to form a Junior Achievement company. With the help of an advisor, your group has lined up some work contracts. You will set up assembly lines to complete special work assignments. Your first contract will be with a fast-food restaurant chain. The company specializes in fried chicken, french fries, and soft drinks. With each meal, the counterperson at the fast-food restaurant gives the customer a table-setting packet. This packet comes in a sealed plastic bag. Within the bag are a plastic fork, knife and spoon. Also in the bag are a paper napkin and a moist towel wrapped in foil.

**Teamwork is essential** in technology, just as it is in sports. (Courtesy National Basketball Association)

Your contract is to pack these materials and deliver finished packets to the supply center of the fast-food chain. Your customer will have all of the materials shipped to the home of one of the members of your group. The fast-food chain will also provide a machine to seal the bags after they have been filled. Your team will set up tables in the den of your friend's home. Work stations will be set up to create an assembly line. At each work station, another article will be put into a bag. When the bag is filled, it will be sealed and placed in a box for delivery.

Within your assembly-line system, the input function involves setting up your work stations. This includes providing supplies of materials for each person who will put an article into a packet.

The processing function is the actual stuffing of the supplies into the packets.

Output is the placing of filled packets into boxes and delivery of the boxes to the customer.

## Monitoring and Feedback

The real key to making all this work lies in monitoring and *feedback.* Monitoring includes inspection of the finished packets. Someone should look at every packet to make sure that all the articles are included. The inspector should also check to make sure that the plastic bags are sealed properly.

Monitoring also involves gathering feedback about how the operation is going. Feedback, in this sense, is information on

**Individual operations** must be monitored and controlled independently within manufacturing systems. This photo shows the monitoring of the manufacture of microcircuit chips. (Courtesy Sperry Corporation)

what is happening. The feedback can indicate that everything is running as planned. This means that the results that are being delivered are as expected; things are normal. If operations are normal, no corrections or adjustments are necessary.

However, in any complex system, there may be problems at any time. A situation or occurrence that is not normal is known as an *exception*. An exception, in turn, requires attention or correction. Consider some exceptions that could occur in the packet-stuffing operation:

- Workers may complete their operations at different speeds. You might find a large pile of unfinished work at one station. At other stations, workers could find themselves with not enough to do.
- The monitoring function must check supplies of items to be stuffed in bags. This must be done continually. If you run short on a one item, the whole operation may have to close down.
- Quality inspections are vital. All items must be checked as they are received. Suppose you receive a shipment of moist

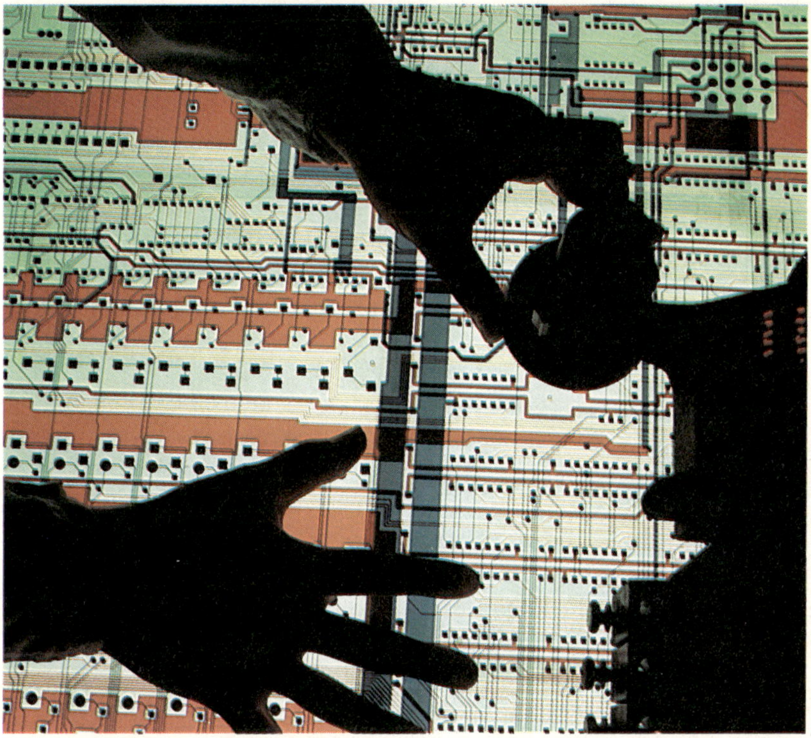

**Quality inspections are vital** at every step in the development and production of technological products. This photo shows inspection of photographic masters to be used in the manufacture of microchips. (Courtesy Storage Technology Corporation © 1983)

towels for which the foil was not sealed properly. You find the towels to be dry and useless to the customers. You have to advise your customer of the problem. You also have to get a new supply of moist towels to keep your operation running.

- The person who inspects the finished packets must check for mistakes. Workers may be putting items in backwards. That is, the fast-food operation wants the fork, knife, spoon all to point the same way. If you don't train your workers properly, items may be inserted backwards. This means expensive rework. It also identifies a need for correction: You have to be sure that all workers know the proper way to complete their jobs.

To make any system work, quality must be monitored. Then, when monitoring uncovers problems, the team must be ready to act. Corrections or adjustments must be made as soon as possible. Without prompt action, errors or losses simply continue. Results of improper processing are unusable outputs. Instead of valuable products, the system may be producing rejects.

## Relationships Among People

The people who work within operational systems have responsibilities to one another. It isn't acceptable for people simply to perform their own operations without regard to others. To illustrate, the packet-stuffing job uses an *assembly line* system. An assembly line consists of a series of people and functions that follow in sequence. Assembly lines are used commonly as manufacturing methods. Products such as automobiles or refrigerators are carried down lines of work stations. At each

**Assembly lines** require a number of work stations and people who perform a series of operations, in order. Workers on assembly lines should be aware that they are a team. Quality products require high performance from all team members. (Photo, courtesy of Hewlett-Packard Company)

work station, people or equipment perform one special operation.

Within a system of this type, each person depends on the performance of those in earlier positions along the line. For example, in the packet-stuffing system, suppose the person inserting knives works slower than the rest. This means that people in later positions on the line don't have enough work to keep going. The whole operation could fail to meet its goals because of failure at any one position.

**Team membership** can extend to all people involved on the same communication link within a computer system. The quality of the data and information depend on the performance of all team members.

Everybody who works within a system has responsibilities to the other workers. One way to look at it is that each person in an operation is a customer. This individual is a customer for all those who have performed earlier operations. Also, the persons performing earlier operations are suppliers to those whose operations follow. When customer satisfaction is present throughout the system, quality output results.

People make systems. People are necessary to operate systems. People must cooperate to produce expected, needed outputs.

## DEVELOPING TECHNOLOGICAL SYSTEMS

The work of systems development uses a specialized technology of its own. Systems are developed through organized methods involving complex teamwork. These efforts are called *projects.* A project consists of a series of steps followed to produce known end results.

In effect, a project establishes a known method for dealing with unknowns. The project steps themselves follow a known sequence. But the development of any given system may require that the project team deal with unknown elements. An organized project approach makes it possible to analyze and solve problems involving such unknown elements.

Space travel provides an excellent example of how projects that produce known results must deal with unknowns. The team that designed the space shuttle mission knew the results. One main purpose was to develop a space vehicle that could be used over and over. This vehicle was to be launched, put into orbit, to re-enter the Earth's atmosphere, and to land like an airplane.

The scientists and technicians assigned to the project team knew that the job could be done. However, many specific parts of the project involved unknown elements. For example, the space shuttle has to fly like an airplane after it re-enters the earth's atmosphere. This had never been done before. Therefore, there was no specific information available about the design of the wings, the shape of the airframe (body) or the landing gear.

**New challenges** are faced at every stage of the development of new technologies. This photo illustrates a special solution for a unique problem—the return of space shuttles to their launchpads. The shuttle vehicles are carried piggyback by specially fitted 747 jet aircraft.

The space shuttle was to carry as many as seven people, who would live in space for long periods. There were many unknowns about what kinds of living quarters and life-support systems would be needed. Had all of these unknown factors been lumped together, the space shuttle could never have been developed. Instead, the project method was used. The overall set of unknowns was broken into a series of small parts. These parts were small enough that experts in each field could deal with them. Detailed study and design made it possible for specialists to solve all the problems. When all of these parts were brought together, the space shuttle became a working system.

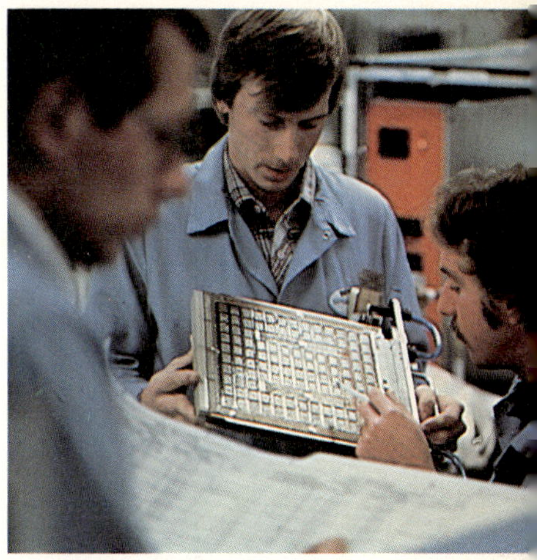

**Problem analysis can be a team effort** when high-technology products are involved. (Courtesy Sperry Corporation)

The breaking down of jobs within an overall project requires special methods. The need is to find a way to keep track of what is happening in a large project. People responsible for the project must know how every group of specialists is doing. Project managers must know where team members have had successes and where they need help.

The methodology for maintaining this kind of control is known as a *project structure*. The project structure involves a series of steps. These steps are used to organize the project, complete technological designs, and produce the known, expected results. By setting up a series of steps, also called *phases*, it becomes possible to manage large technological development efforts. This, basically, is the systems approach. The systems methodology makes complex systems of technology possible. The systems approach ties multiple technologies together. There are many types of project structures. However, a typical set of steps, or phases, that form the project structure is as follows:

- Set goals
- Evaluate feasibility
- Establish a plan of work
- General design
- Technical design
- Development
- Implementation
- Evaluation.

These phases are tied together through a continuing effort that involves project management and information feedback. Feedback is a term meaning a return of results. Feedback is information about how things are going. In project management, feedback is the basis for making decisions and directing the efforts of team members.

The relationship of management and feedback to the phases of a project structure is illustrated in the accompanying diagram.

**Project monitoring** often requires the aid of computers. Computer terminals have become standard tools in factories that make jet aircraft. (Photo, courtesy of Hewlett-Packard Company)

A diagram of this type, that shows the sequence of activities and relationships among activities, is known as a *model.* A model of the kind shown is often used to help plan a project.

Project management is a complex business. It is impossible to explain all of the elements of project management in detail within this book. However, you live in a technological world. You will eventually be expected to work and survive in this technological world. Therefore, as a working citizen of the future, you should know something about systems of technology and how they are brought into existence. The descriptions that follow provide a basic introduction that should be helpful.

## SET GOALS

Beginning a project is something like starting a long trip through strange areas. Before you start, you should find out where you are going—pick a destination. If this is your first visit to an area, you will want to look at a map before you start.

Project-type studies look well into the future. Of necessity, many projects deal with information that team members have never encountered before. Some examples:

- A project team might be planning a company's future. The job of team members is to determine what the company should look like in the future. Team predictions may look ahead five or seven years. Of course, nobody can say for certain what will happen in five or seven years. But it is possible to look ahead and think about some things that might happen. The team can then plan to deal with situations when and if they arise. Planning is a project activity. The more effective the planning, the better prepared an organization will be for future opportunities.
- A project that leads to the development of a new car, airplane, or computer might run between three and five years. The products are complex. Dealing with all the unknowns involved—making all of the needed decisions—can take years. Hundreds of people can be involved.
- Many projects are aimed at developing technologies to protect or improve human health. A project of this type might start with a goal of finding the cause of a disease. The systems approach represents an important way to deal with the many unknowns faced by people in biotechnology.

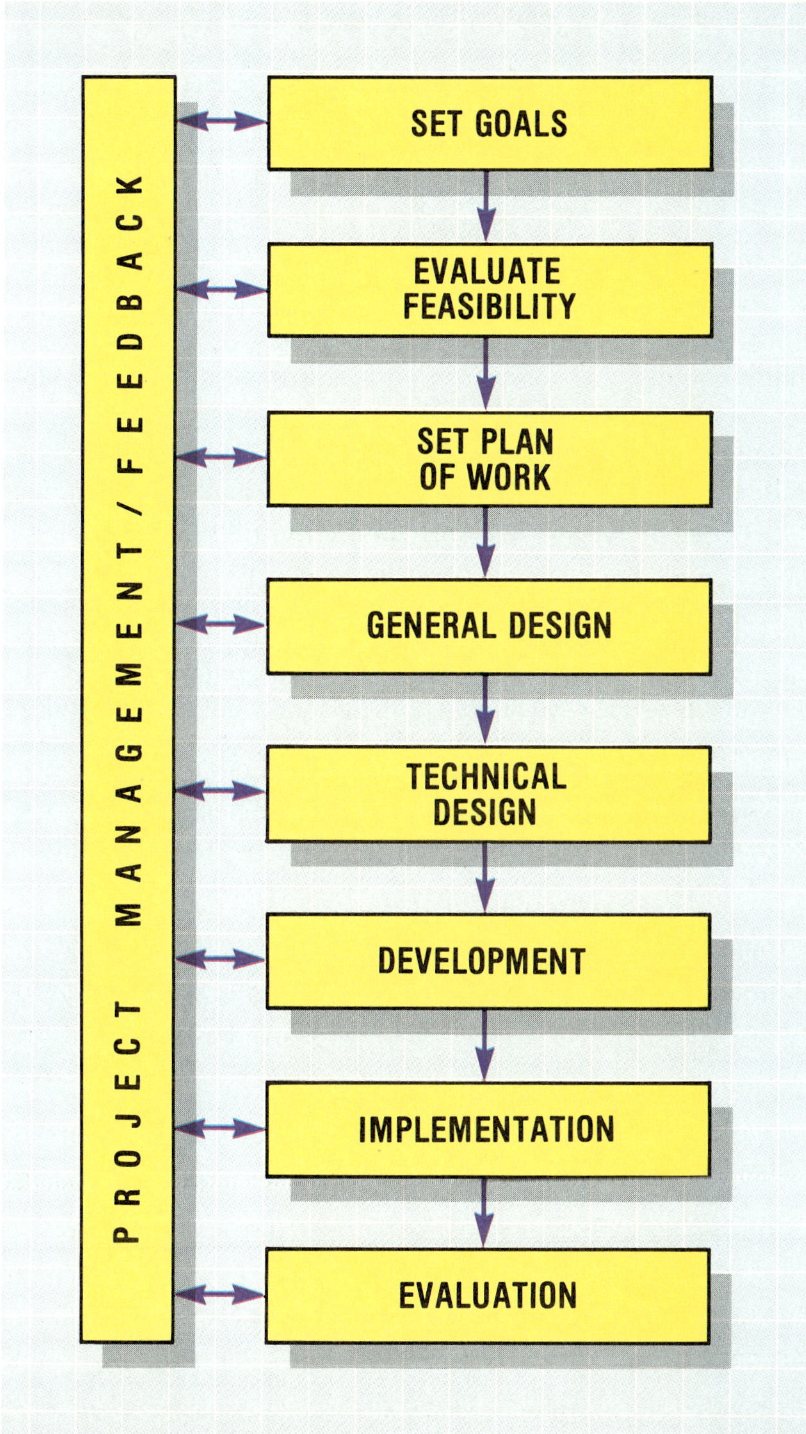


**The model for the project structure** used to develop systems of technology includes a series of orderly steps. There is also continuing project management and feedback of information.

In determining the goals for a project, the team draws boundaries. That is, the statements that set goals determine what is to be included in the project. Just as important, the project goals should indicate areas that are excluded. No individual study project can take in all of technology. Each project team must set boundaries within which it will work.

## A Personal Project

Most projects involve many people and take quite a bit of time. However, the project approach can work for any complex situation involving a move into the future, through unknown areas.

To personalize the project structure, consider a situation in which you will be involved. This project deals with career selection. You are to use the project approach to look into your own future. The goal is, in effect, to design the adult that you would like to become.

Examples of how this project would work are included in the rest of this chapter. Work through the project yourself. You may find help in developing some plans for your own future. At the very least, the examples will help you build your understanding of the systems approach and of project management.

In a personal study, your goal might be to develop a picture of yourself at some future time. A goal might be to design a future for yourself. This future includes the work you will do and the life you will live at age 25.

This is an ideal situation for a project approach. There are certainly many unknowns for you between now and age 25. You might even decide that you can't develop an exact picture. However, any decisions you can make or goals you can set will help. The more you know about your future self, the better prepared you will be for the future. You will prepare yourself to live with that new person you are in the process of becoming.

**Project management methods** can also be applied to personal problems or projects. To start such an effort, simply think ahead and set goals.

At the personal level, getting started on a project need not be complicated. All you need to know now is that you want to look ahead, to set a course.

On a larger project, this first phase, or step, would be more complex. Statements of goals and directions can take up dozens, even hundreds, of pages. The more complex the project, the more people there will be on the project team. For a difficult project involving many people, goal setting can be a big job. On formal projects, it may take dozens, even hundreds, of pages to list the goals to be met.

**Project goals** must be understood clearly if people are to work together toward the same end. (© Liane Enkelis)

## EVALUATE FEASIBILITY

In setting goals, you define the job to be done. Now, in the second step of a project structure, you determine whether the job you have defined can, in fact, be done. One way of doing this is to ask a series of questions:

- Can the goals that have been set be achieved?
- If the goals can be achieved, are they really worth going after?
- What are the benefits of success?
- What are the costs of failure?
- Is this project a good risk?

In effect, these questions lead you through a series of decision steps. The first is to determine whether the job can be done. The second involves a decision about whether the job is worth doing. After that, the questions lead you to identify the benefits that can be realized and the costs that will be involved. Finally, there is the question: "Is this project a good risk?"

The answer to the last question becomes a decision. In effect, then, this second step is a checkpoint. The activities involved lead to a decision about whether the project is worth doing. Every project involving technology requires effort and involves potential benefits and costs. A major project can involve costs of millions, even billions, of dollars.

One example of advanced, expensive technology involves the field of robotics. A company may be considering building an automated factory. Costs may run into hundreds of millions of dollars. Benefits can include higher quality and lower costs

**Quality checkpoints** must be incorporated in all products applying technolgies. Establishing quality checkpoints is part of project planning. (Courtesy Sperry Corporation)

for the manufacture of products. Many companies that make or use microchips or small computers face such decisions. Chip making, and even computer assembly can be done with highly automatic equipment. The manufacturing equipment is far more expensive than the units used for hand assembly. But labor costs are lower.

In effect, then, a company undertaking a project of this type faces a trade-off. On the one hand, money must be spent for the new factory and equipment. On the other hand, each unit of equipment produced in the factory costs less to make. The reduced costs are the benefits. The expense of the factory is the major cost. To determine whether the benefits or the costs are greater, project team members must look to the future. They

**Microscopic detail** is sometimes needed in high-technology work. The bright light in this photo makes it possible to perform microscopic examinations of microchips during manufacture. (Courtesy Sperry Corporation)

must decide how many of the manufactured items can be sold if the plant is built. The best guess about the number of units to be sold holds the key to a decision.

As a real-life example, consider what happened to Chrysler Corporation between 1979 and 1984. The company's cars had not been selling well. The company had lost money on its operations. Chrysler was close to bankruptcy—financial failure.

In this situation, management of the company borrowed a large amount of money. The money, in turn, was invested in improving factories and in designing new cars. The new-car designs used advanced technologies. Company management decided that it was worthwhile to invest in the plant and equipment to build the new cars. By 1983, Chrysler Corporation was profitable again. By 1984, the company was realizing the greatest profits in its history.

However, not all investments in technology are automatically successful. There are many instances in which companies have invested money in building products that consumers didn't want. Failures have resulted.

## Your Personal Project

Deciding feasibility about a look at your future presents no major problems. You can count many benefits. First, you will get to know yourself better. Second, if you are successful, it will bring added meaning to your school and work experiences. You will

**Self-assurance** can help make for success in the world of technology. It is important for an individual to know himself or herself and to set goals. Goal setting can be an important part of a personal plan of work.

give yourself a reason for the classes you take or the jobs you will apply for. Life can be more interesting if you have a sense of direction.

There are no major costs involved. Thus, the potential benefits clearly outweigh the costs. There is no risk. The situation obviously calls for a decision to go ahead with your personal project.

## ESTABLISH A PLAN OF WORK

A *plan of work* applies to the project itself. The plan covers the work to be done, step by step, to achieve the stated goals of the project. The plan of work does not actually deal with the system of technology to be developed. This an important distinction! Before a project can deal with actual technologies, the people involved must know what is expected of them. The plan helps the project team to get organized for the work ahead. No actual technological effort is covered by the plan.

Among the important results of this project step is the setting of *checkpoints* during the project. A checkpoint is a pause in a project to review what has happened. At a checkpoint, members of the project team review their goals. If necessary, the plan of work is adjusted to match conditions that have been found.

Among the elements that make up a plan of work are a series of documents that:

- Identify all jobs, or tasks, to be performed throughout the project.
- Identify the people who will work on the project.
- Match the skills of the people to the requirements of the jobs or tasks.
- Schedule all work to be done. Depending on the length of the project, the schedule can be on a weekly or monthly basis.
- Establish a *budget* for the project. A budget is a plan for spending money to produce results.
- Set up a method for monitoring progress on the project. The monitoring technique looks at work completed and compares actual performance with schedules and budgets. In other words, the monitoring system manages the project.

Persons responsible for monitoring check what has been completed. Then they compare the work that has been done with the schedule that shows what should have been done. Work completed is also compared with the budget that shows what should have been spent.

Technology is part of an area of knowledge known as applied science, or *applied technology*. In this sense, applied means that principles and theories of science are used to produce specific results. Considerable amounts of money are invested in projects aimed at developing systems of technology. It makes sense to establish a plan of work and to create controls to make sure that the desired work gets done.

## Your Plan of Work

Now consider a plan of work for your personal project aimed at looking toward your future. You would begin by identifying the elements of your adulthood that should be studied. You have some important decisions to make. These would center upon the life-style you expect, your level of education, and your choice of career. Life-style refers partly to where and how you live. But considerations go further. You must think about how you expect to dress, the kind of neighborhood or area you want to live in, and what you expect to earn. Your general decisions about life-style might affect the kind of education you plan for. Certainly, your career decisions will relate to both education and life-style.

**Knowing yourself** is an important step toward the setting of personal goals. Self-examination should be as objective as possible. Knowing yourself now helps you to determine what you feel you can become.

In effect, these three areas of study—life-style, education, and career—develop a picture of how you think you will look when you are 25. Your plan of work for developing this picture will begin with you. To look into your future, you will have to get to know the present-day you quite well. Think seriously about yourself. Ask yourself what you expect out of life. Then ask yourself if you are able to meet your own expectations. That is, you want to make sure that your expectations about life-style, education, and career are realistic.

In addition to your expectations, you have to be honest with yourself in projecting what you think you can do. In other words, study and understand your own strengths and weaknesses. Strengths are things that you do well and enjoy doing. Weaknesses are things that you don't do well or don't enjoy. If you don't like doing something, it doesn't make sense to plan on doing that thing for the rest of your life.

The way to point yourself toward future success and satisfaction is to match yourself and your opportunities. Gather information about life-styles, about costs of living, about education and careers. Then match this information against the picture you have developed about yourself. What career and life-style combinations meet your expectations best? If you have a match, will your strengths and weaknesses enable you to achieve those goals?

**Realizing expectations** is an important part of job satisfaction. In setting expectations, a person should be aware of his or her own strengths and weaknesses.

Your own plan of work, then, is to study yourself and your opportunities. Then you want to match the person you will become with the opportunities you have chosen.

## GENERAL DESIGN

Note that there are two design steps, or phases, in the project structure. The first step, general design, identifies the results to be delivered. At this point, the project team does not get deeply into applied technologies. Instead, emphasis is upon identifying the human wants and needs to be met.

To illustrate, suppose the project team is designing a new airliner. Members of the team would use their knowledge about airline companies and passengers as a basis for this phase. One set of decisions to be made, for example, would center upon performance of the airplane. Team members would decide how far the plane was to fly and how many passengers it should carry. Other factors would include arrangements for passenger comfort and feeding.

Such considerations establish general expectations for the performance of the airplane. Notice that there is no mention of such factors as what engines to be used. Neither does work in this phase deal with design of the airframe, the landing gear, or the electrical system.

**Team members should meet** during the course of a project. The idea is for everyone to know what is happening. Understanding is a key to the general design for a product or system of technologies.

General design, then, develops a picture that presents the product as its users will see it. A passenger doesn't see the engines or landing gear of an aircraft. These are concerns for engineers and other technicians. Passenger decisions about an airplane center around such factors as personal comfort and cost of travel. So, these considerations are covered during the general design phase.

**Life-style expectations** should be an important part of an individual's personal planning.

## Your Personal Project

At this point in your personal study about your future, concentrate on identifying the life-style you expect. Ask yourself what you expect from life. Determine the things you consider to be necessities. Also identify your wants. Develop a picture of how you expect to live and what you expect to do as a person.

You don't have to worry about how you are going to achieve the life-style you identify. That consideration comes during the next phase. You can always modify or compromise your ideals. During general design, you should develop an ideal picture.

## TECHNICAL DESIGN

This phase of a project concentrates on how to build the product or achieve the results defined during general design. This is the point in the project when engineers and technicians roll up their sleeves and go to work. The efforts are highly technical. In designing an airplane, for example, technical teams work on the airframe (body), landing gear, and flight control systems. Hundreds of highly trained technicians and engineers may be involved.

An important feature of technical design is that results sometimes challenge the general design. That is, features included in the general design may not work technically. If so, consideration is given to changing the general design. The process is known as feedback. That is, information is fed back to earlier points in the project. Based on findings, designs may be

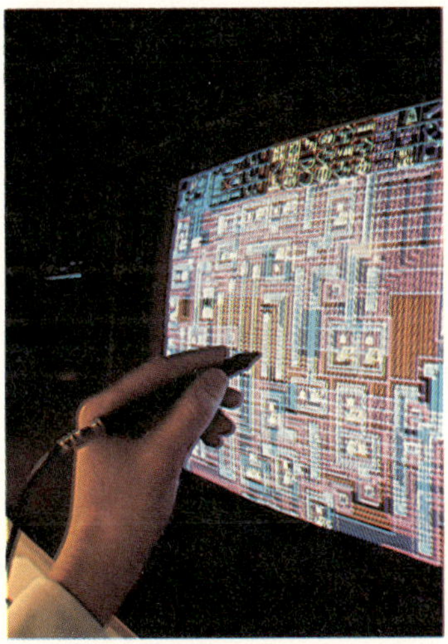

modified. Another alternative might be to use a different technology. Also, if technical design efforts prove that a general design won't work, the project may be abandoned.

**The technical design phase** of a technology project involves considerably more detail than the general design phase. Examples can include the setting up of electronic circuits or the testing of aircraft in wind tunnels. (Both photos, courtesy TRW, Inc.)

### Your Personal Project

At this point, you find out if the ideal picture of yourself can be achieved. You learn about the costs of food, housing, clothing, and entertainment. You learn about career opportunities and how much different types of jobs pay. Then you match prospective earnings with the costs of the life-style you have selected.

Suppose you enjoy music in your home. You would decide about the quality of the stereo equipment you would expect to own. Then you would find out how much this equipment costs. Relate this cost, and others, to your ability to pay for the things you would like.

At this point, you may find that you have to adjust your life-style targets or change your career goals. In either case, the learning process is valuable. Once you match your career goals and your life-style expectations, you are on your way to a satisfying future.

## DEVELOPMENT

In the development phase, the project team gets set to build the product or deliver the planned result. Activities can include setting up a factory, ordering machinery, purchasing materials, and training people.

As development takes place, additional feedback is created. It may be possible, for example, that a technical design is hard to build. If this situation occurs, some last-minute adjustments to technical designs may be needed.

Once a product is in development, a major commitment has been made. Most of the budgeted money is spent during development. At this point, the company may even be working out an advertising and sales campaign for the new product. Once development starts, the organization is committed to the technology that has been selected.

**The development phase** of a technology project calls for the actual building of equipment. Very often, this will involve electrical or electronic work. (Courtesy Wang Laboratories, Inc.)

Modern electronic systems are so complex that the need for a product development process becomes obvious. (Courtesy TRW, Inc.)

## Your Personal Project

At a personal level, the development phase of your project would be devoted to planning. You would decide on a course of study to follow. In other words, you are beginning to make yourself into the person you want to become. You may also want to plan for after-school or summer jobs. These may help lead toward your future in two ways. First, the money you earn may help finance your continued development. Second, the experience you gain may lead toward personal growth.

## IMPLEMENTATION

To *implement* is to make something happen. During the implementation phase of a project, the finished product becomes a reality. Steps involved can include shipping, advertising, and selling the new product. In terms of technology, these steps may be less complex than technical design and development. However, in the modern world, distributing and marketing products often costs more than making them. Until the program is implemented, there are no real results. So, the implementation step is critically important.

## Your Personal Project

Every time you take a step leading toward a planned goal, you are implementing part of your future. Be aware of your progress. Measure your satisfaction at each major step.

**Final implementation** of a project can involve the creation of large manufacturing facilities. (Courtesy TRW, Inc.)

## EVALUATION

An old saying holds that experience is a great teacher. The evaluation phase of a project is aimed at learning from experience. At this point, the development program has been implemented. Products are being delivered and sold. An important step at the close of any project is to evaluate results and to learn from experience.

If a technological project is involved, evaluation may lead to *modification*. Modification means change, usually for improvement. Companies are always finding small changes to make in existing products. These changes improve the products or reduce the costs of making them. The feedback from evaluation should make for continuing progress.

**Evaluation of results** can involve extensive technical analysis of the work in a project. On a personal level, use of the project approach should enable individuals to step forward toward their futures with confidence. (Photo at left, Storage Technology Corporation © 1983)

## Your Personal Project

Each time you implement a step toward your future, evaluate your progress. *Orient* yourself for the next step. To orient is to find out where you are and where you are going. Throughout your life, you will have opportunities to change or redirect your goals and efforts. Planned evaluation will help you to become the person you want to be.

## MANAGEMENT AND FEEDBACK

Feedback is essential throughout a project. Feedback presents opportunities to monitor progress and to make any necessary adjustments. This is what management of technology is all about. The idea is that technology should serve people. Management controls the development and use of technologies, making sure that technologies serve people.

The systems approach is a management method. By using the systems approach, people assure that they will realize the results they expect from technology. This applies on all projects for development of technology—personal or organizational.

## Summary

Pure science is different from applied science. Pure science seeks to build knowledge through exploration. Applied science, on the other hand, results in technology that meets people's needs and wants.

Operation of a system that applies technology follows a method similar to that used for information systems. That is, the system has separate input, processing, and output functions. Other functions include supplies to be used and a monitoring of all operations. The monitoring function makes adjustments or changes when problems arise within the system.

Because modern technology has become complex, new technologies are developed through a systems approach. A systems approach, or systems development, coordinates the many elements that go into creating systems of technology. For example, an automobile is made up of many systems of technologies: energy, electricity, combustion, structure, lubrication, etc. The complexities of modern technologies require a management method to organize efforts and control details.

This management method is a systems approach. First, a project is outlined. That is, team members set the goals of a particular project. Then, the feasibility of the project is evaluated. This means that people decide if a project can be done. Next, a plan of work is devised. A plan helps team members to know what needs to be done when. Also, budgets are set during this phase of a systems approach.

Once the plan of work is established, general design is begun. General design develops the picture of what the final product will be. After the general design is set, technical design takes over. In this phase, hundreds of highly trained technicians and engineers may be involved.

Once design of the end product is settled, actual development begins. This includes activities such as setting up a factory, ordering machinery, purchasing materials, and training people. After the product is available, implementation can take place. Implementation includes shipping, advertising, and selling the product.

As with any system, if any one step of the systems approach fails, the system won't work. Implementation, while it may appear less important, is as necessary as technical design.

Feedback is a recurring element throughout all phases of a systems development. Feedback presents opportunities to monitor progress and make necessary adjustments.

The systems approach is a management method that can be applied to any project, whether in a personal or a technological sense.

## Technology Terms

pure science
applied science
systems approach
systems development
hydraulics
pneumatics
computer architect
model
monitor
feedback
exception
assembly line
project
project structure
phase
plan of work
checkpoint
budget
applied technology
implement
modification
orient

## Using Technology

1. For this activity, you will work within a group. Suppose that you have been told that a vacant lot near school might be available for a student-organized event. Suppose that your group is elected to explore this opportunity. Use the systems approach to develop your project. Record the steps that your group would take, from setting the goals to implementation. After your project has been developed, share your reports with the class.

2. In a short composition, explain why a systems approach is necessary in developing modern technologies.

3. Automobiles, computers, and airplanes were described as examples of systems of technologies. Can you think of others? Jot down your examples to share with the class.

# 16

# TECHNOLOGY AT WORK

### Your Learning Job

After you have read this chapter and completed the work assignments, you should be able to:

- ☐ State briefly the role of technology in agriculture.
- ☐ Outline the technologies involved in the area of infrastructure.
- ☐ Describe a modern warehouse.
- ☐ Describe, in a general way, how the military and aerospace fields are involved with technologies.
- ☐ Explain the difference between the two major areas of communication: transmission and information.
- ☐ Describe the ways in which a bank or other financial institution uses technology.
- ☐ Outline the effect of computer technology on education.
- ☐ Describe the ways in which computers are affecting offices.

New opportunities in a world of technology are illustrated by the workers at this computer control center. The terminals in this photo display information about the operation of several large computer systems. (© Liane Enkelis)

## TECHNOLOGY AND OPPORTUNITIES

Technology meets needs. For a technology to be accepted by people, needs must exist and be understood. These needs must exist among people as individuals, among organizations of people, or in entire segments of society. Remember, people are the potential users of new technologies. Meeting the needs and wants of these potential users of technology, in turn, leads to opportunities. If enough people need or want a technological product, that technology becomes a potential career opportunity. That is, there are careers to be developed in any area that can be identified as a major user of a given technology.

As you look ahead at the person you are to become, you should examine existing and future technologies. In your future study of technology, you will review the opportunities connected with each of these areas. Now, at the close of this book, you can take a first step. The discussions that follow cover some important areas of technology at work. For each area, there is a brief discussion of the role of technology. Also reviewed are the potential opportunities for people who prepare themselves to work in each field. Technologies covered in the remainder of this chapter include:

- Agriculture
- Infrastructure
- Production
- Materials handling and transportation
- Aerospace/military
- Communication-transmission
- Communication-information
- Commerce and banking
- Education
- Administration.

In addition, the chapter concludes with a discussion of the effects of technology on personal services and life-styles.

## AGRICULTURE

The field of agriculture has felt the effects of technology more than any other occupation. At one time, America was looked on as a nation of farmers. Some 70 percent of the people were involved in farming or food processing. Today, that figure is about 3 percent. You have already read, in earlier chapters, about equipment for planting and harvesting crops. These technologies have made the difference in reducing requirements for human labor in agriculture.

**Agriculture** has become a heavily mechanized business. Knowledge of technology can help lead to success in this field. (USDA Photo)

**Special agricultural techniques** have made it possible to farm hillsides and other areas that were not suitable years ago. Technology is expanding the potential in agriculture. (Sperry-New Holland)

You also know that many farm managers and people in other branches of agritechnology use computers. Agriculture has become a business in which success lies in the management of resources. The main resources are land and weather. If the weather doesn't provide enough water to grow crops, farmers need irrigation technology. Decisions about which crops to grow and what units of land to use are complex. Computers are used commonly to help farmers plan their operations. Computer programs help predict costs for seed and fertilizer. Computer programs also help to plan for income from the crops to be grown. Computers even help farmers to decide when to buy new planting or harvesting equipment. Computer programs indicate how new technology will affect a farmer's income.

As you know, the broad field of agritechnology takes in the handling and packing of farm products. This field offers many opportunities. There is a saying among people in the food field: "No matter how bad things get, people always have to eat." This means that people who have a solid position in agritechnology can expect a continuing demand for their products and services.

Farming and other branches of agriculture can be satisfying. Many people experience joy at planting crops and seeing them grow to produce food. If you are among these people, just remember that you will need some special skills. Modern agriculture is a high-technology business.

## INFRASTRUCTURE

*Infrastructure* is a broad area involving many technologies. The term infrastructure refers to all of the facilities that people in societies need for their support. Included are roads, airports, railroad stations, office buildings, factories, and schools. Infrastructure also takes in sewers and waste disposal systems, as well as supplies of water, gas, and electricity. The telephone system is also part of infrastructure.

**Infrastructure development** will be a continuing challenge well into the future. Many opportunities should open in this area. These include construction of new buildings. Also involved will be the construction of special aqueducts to bring water to cities. (Photo at left, Queen City Metro, Cincinnati, Ohio; photo at right, courtesy Metropolitan Water District of Southern California)

In general, infrastructure involves construction. Construction, in turn, depends upon many technologies. One type of technology involves planning for the use of land. This planning includes decisions about where parks, roads, and buildings will be located. Actual design and construction of the infrastructure facilities are handled by many businesses and industries. Construction work employs many people. Construction is one of the basic industries of a modern society.

People drawn to jobs involving planning and construction tend to be good at visualizing the results of plans or ideas. To visualize means that you can see in your mind what something will look like. An architect, for example, can look at a hillside or an empty field and visualize its end uses. The architect may envision a factory, a shopping center, or a group of homes. A skilled builder can look at a series of construction drawings and visualize a finished building.

People who work in construction also enjoy being out of doors. They get satisfaction from doing physical work.

Construction is a high-technology field that is changing rapidly as new materials and ideas are introduced.

## PRODUCTION

The field of production, or manufacturing, provided one of the birthplaces for modern technology. The Industrial Revolution concentrated on improving production. Production of goods was moved from homes to factories.

During the early part of this century, manufacturing went through a kind of technological revolution. The change came with the introduction of mass production techniques. Henry Ford and others introduced the assembly line during the early

**Factory operations** will depend increasingly upon computers. This will apply both to computer-produced documents and also to direct computer operation of equipment, as illustrated on the next page. (© Liane Enkelis)

**Automatic manufacturing equipment** should come into increasing use in the future. Persons who can build or control these devices will enjoy excellent job prospects. The photo above shows a robot welding machine. At right is a 'clean room' in which computer parts are made. (Photo at left, courtesy Cincinnati Milacron. Photo at right, courtesy Sperry Corporation.)

twentieth century. Production jobs became a main source of income for working people.

Today, production is going through a new type of revolution. Production is becoming one of the front lines of modern technology. Electronic and mechanical technologies are being combined to provide automatic manufacturing. The field of robotics is growing rapidly. As robotics techniques are accepted, manufacturing techniques change drastically.

In the future, the field of production, or manufacturing, will provide fewer jobs. In addition, the jobs that will exist will tend to require higher levels of skills. Opportunities will develop for people who build their skills in such areas as electronics or mechanical engineering.

## MATERIALS HANDLING AND TRANSPORTATION

Moving goods and people has become a big job. Materials handling and transportation activities involve a major series of industries using a combination of technologies.

The United States has been a pioneer in mass transportation. This pioneering effort has been needed largely because of the size of the country. As the United States conquered its frontiers, the country found itself with a distance of 3,000 miles between its shores. This represented a different dimension in travel for the people from Europe who had settled and built the United States. In Europe, you can travel through a dozen or more countries within the time and distance that it takes to cross just a few states in America.

**Transportation equipment** relied on steam for power for some two centuries. This century has seen a transportation revolution that will continue to create new opportunities well into the future.

During the nineteenth century, the United States pioneered new forms of transportation technology. The United States led the world in building railroads. American companies were leaders in building and using steam-powered ships. The airplane, which has become the main form of long-distance transportation for people, was invented in the United States.

During the twentieth century, transportation systems have expanded rapidly. These systems now include thousands of miles of highways, built at costs of billions of dollars. With the growth of highway systems, automobiles have become the main means of personal transportation. Trucks have become the main carriers of the goods and supplies that people need.

This group of technologies also includes the handling of materials in places where they are made or stored. This field is known as *materials handling*. At one time, most materials handling work was done by people. It was simply a matter of picking things up, carrying them to where they were needed, then setting them down. In the past few decades, materials handling has become highly mechanized. Complex conveyors are used to carry materials from one location to another. Automated devices pick up, store, find, and deliver materials.

**A modern system of highways** assures convenient transportation. In turn, jobs are created in the building and repairing of roads, as well as in the building and maintenance of automobiles.

There are *warehouses* in which people no longer have to handle the goods that are stored. A warehouse is a facility designed to store equipment or products. Many modern warehouses are computer controlled. The computers direct the operation of equipment that handles full cartons of products. The equipment picks up the products, delivers them to storage locations, and puts the items on shelves. The same equipment can find needed items and deliver them to trucks for shipment.

A dramatic example of how technology has changed transportation can be seen in the movement of freight by ships. Articles to be sent across oceans used to be loaded one at a time by people. Today, much of the cargo carried on ocean-going

**Air shipment** of manufactured goods is becoming an increasingly important means of transportation. Airplanes are making rapid shipment and delivery of goods possible and affordable.
(U.S. Customs Service)

ships is loaded and handled in containers. Entire truck-sized boxes are loaded by the shippers. These complete containers are picked up and placed aboard ship, greatly reducing the labor involved.

The materials handling and transportation field offers a good example of how transportation is changing labor markets. Because of automatic equipment, there are fewer jobs that involve simply picking up and loading boxes of materials. Instead, there are increasing numbers of jobs for people who have the skills needed to operate special machinery. Also, the design and building of materials handling equipment represents a technology area that should continue to grow.

## AEROSPACE/MILITARY

Most of the people and organizations in the aerospace and military fields serve a single customer. This customer is the United States Department of Defense. The money spent on aerospace and military activities represents the biggest single part of the budget of the federal government. Some $200 billion is spent in this area every year.

Much of the attention given to the aerospace and military fields goes to aircraft, missiles, and other weapons. This focus of attention loses sight of the people in the aerospace/military picture. The biggest item in the aerospace/military budget is for the salaries and benefits paid to military personnel. Millions of people are employed by military organizations. Many of these are civilians handling engineering, administrative, and office jobs. Well over 1 million others are uniformed personnel. These are the soldiers, sailors, marines, and coast guard members responsible for defending the country.

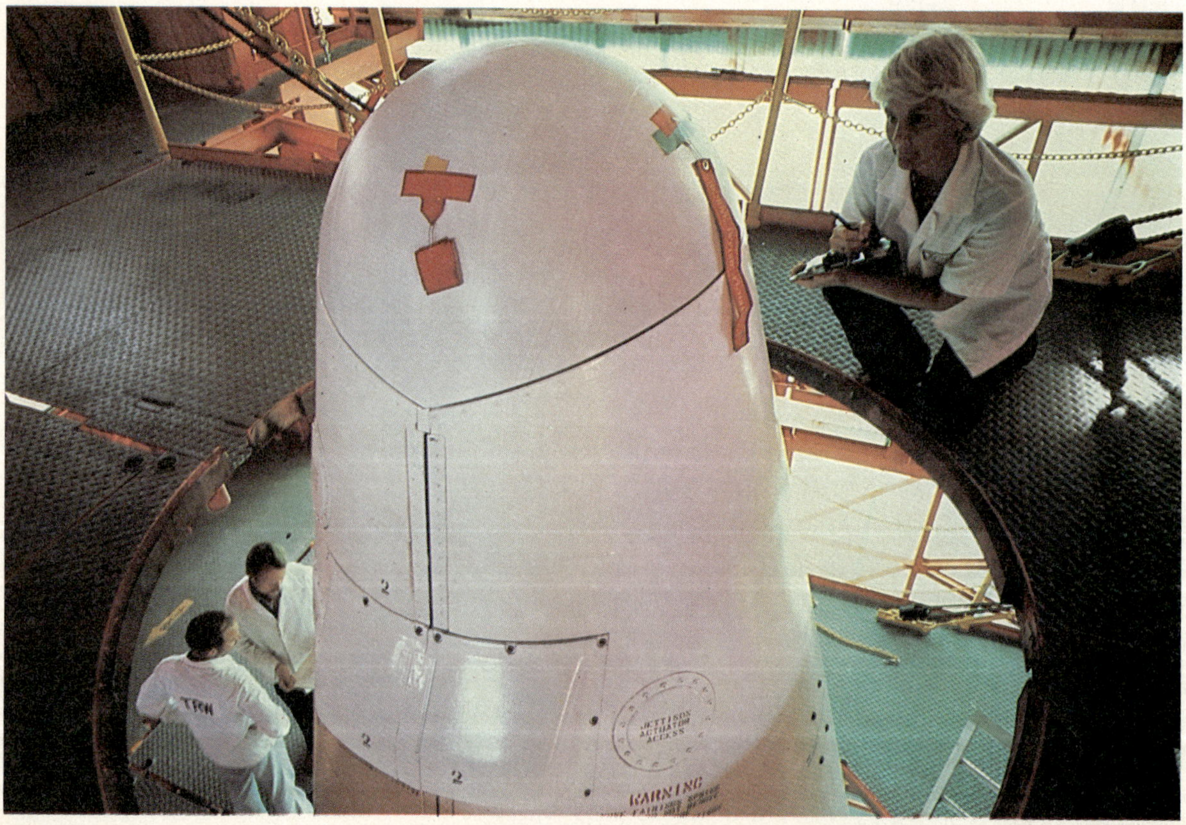


**Aerospace and military** production involves high technology and careful quality control. Continuing progress in these fields means many job opportunities. (Courtesy TRW, Inc.)

It is well known that the aerospace/military area is a leading developer and user of high technologies. Tens of thousands of scientists, engineers, and technicians work on the development of communications and weapons systems for military use. Many of the inventions created in this field find their way into civilian use. Civilian use means that a product is used for the general population. In military terms, a civilian is anyone who is not in a uniformed job.

In addition to military projects, the aerospace industry also handles civilian space exploration projects and the building of commercial aircraft.

Research and product design take place continuously in the aerospace and military fields. There are many opportunities for people with advanced technological knowledge. In addition, uniformed military positions have become excellent opportunity areas for many young people. Persons right out of high school or college often enlist in military organizations for the training they will receive. Military organizations have extensive education and training facilities. Members of the armed forces receive high-technology training as part of their job assignments. In effect, people are paid to go to school. Many of the skills

**High-technology frontiers** are often explored in stages. The small aircraft in this photo is the M2F1 lifting body. The small craft was developed to test the principle that has since been used successfully in the space shuttle. The idea is that part of the lift for the aircraft comes from the shape of its body. (Courtesy NASA)

acquired at these schools and on these jobs have future value for the individuals. For many persons, military experience forms a bridge between an unskilled starting point and careers in highly-skilled fields.

## COMMUNICATION—TRANSMISSION

The broad field of communication has been one of the major technology growth areas of the twentieth century. Consider that before this century started, the telephone had been invented but had not come into general use. There was no commercial radio. Television did not experience major growth until the second half of the century. All of these fields are technology-dependent.

The technologies for transmitting communication signals are still growing rapidly. For many years, emphasis was on stringing wires or cables between cities and under oceans. More recently, *microwave, laser,* and *fiber optics* methods have led to improvements. These are all new technologies used in the transmission of signals. Also, satellites are providing higher-quality, lower-cost methods for long-distance transmission of signals. People who understand and can work with these technologies are assured of employment and opportunities.

**Electronic communication** techniques, represented by this microwave relay tower, are opening new opportunities in the field of communication transmission. (Courtesy Bell Laboratories)

Note that the field of communication has been divided into two parts. The transmission of signals applies technologies of its own. The actual creation of information to be transmitted is a separate field. The information portion of the communication industry uses both human creativity and special technologies of its own.

## COMMUNICATION—INFORMATION

Information, remember, is the basis for knowledge. Information comes from the processing of data. Computers are tools for processing data. However, the main sources of control for and users of data processing technologies are people. People have to decide what information is needed. They have to create the data from which the information is developed.

Information also comes from ideas developed by people. The newspapers you read and the newscasts you hear on radio and television are developed by people. The ability to present information in writing is a skill on which technologies depend. The actual writing comes from people and does not depend on technology. However, the communication of information

**Technologies of tomorrow** include laser beams that transmit information on concentrated light beams, left. At right, a technician works on the production of one of the world's largest computers.
(Photo left, © Gary Gladstone/The Image Bank. Photo above, courtesy of Cray Research, Inc.)

generated by people involves high levels of technology. Computers set type for newspapers. Computers control modern high-speed presses that print newspapers, magazines, and books. Computers also control the operation of radio and television stations that broadcast information. In addition, the printing and broadcasting processes themselves use high levels of technology.

The statement is made above that technology depends upon information created by people. Words and pictures are needed to help people understand technology. Other words and pictures are needed to teach people how to use technological equipment. Thus, people who have skills with words or pictures and an understanding of technology are in great demand.

## COMMERCE AND BANKING

Historically, money evolved as a means of exchange. People who had goods to buy or sell used money as the basis of value. As society developed, however, money took on important new dimensions and roles. Today, money and the handling of money represent a major resource for society.

Computers are important tools for putting money to work. Banks, for example, were leaders in introducing computer processing methods. Other money-centered industries include insurance and finance companies. These organizations, too, have been leaders in the use of computers.

Today, computers are standard tools for almost everybody who works in banking or commercial companies. Even customers of such companies are making increasing use of computers. Automatic teller machines represent an important trend

**Modern banks** use major computer systems like the one pictured here. (© Liane Enkelis)

in computer use. Earlier, computers had been operated only by employees of the organizations that owned them. With the automatic teller machine, self-service has come to the handling of banking transactions. Banks actually dispense money to customers who operate computer terminals on their own.

The banking industry offers a good example of how technologies are developed to meet the needs of people and organizations. By the early 1960s, bank customers were writing some 40 billion checks a year in the United States. These were more pieces of paper than people could handle efficiently. Banking organizations spent many millions of dollars to develop computers that could handle checks automatically. Today, the volume of checks has grown to some 80 billion annually. Banks can no longer operate without computers.

With the availability of computers, banks, insurance companies, and other commercial organizations have become, in effect, paperwork factories. Computers are used in each phase of almost every job within the commercial and banking industries. Operations within these organizations have changed. Computers and other electronically-based tools have created many challenging job opportunities. With computers, these organizations can plan for and use money more efficiently. Remember, money is one of the basic resources of a technological society. Therefore, people who work in banking and other financial fields are important users of technology.

## EDUCATION

Education has dual responsibilities where technology is concerned. First, schools must equip people with the knowledge needed for success in a technological world. Schools must offer courses that help to build technology-related skills. Such knowledge of technology is becoming increasingly necessary for personal and professional growth.

**Education in the future** will involve increasing use of computers.

To achieve this level of learning, schools are increasingly using computers as learning aids. From here on, all students will learn about computers and their operation as part of their basic educational programs. In addition, many students will actually use computers to build their knowledge. Computer programs present information to learners. Then, the computers follow through to make sure that the presented information is understood. Hundreds of thousands of computers are already being

**Computer skills** will become increasingly valuable results of educational programs. Many schools already use computers as learning aids.

used in classrooms. In the future, this computer population will grow into the millions.

## ADMINISTRATION

The field of administration takes in the operation and monitoring of the work of organizations. Administrators are responsible for the management of projects and operations within organizations of every type. Most administrators work in offices. Office workers have been among the last to realize the benefits of technology. However, office automation is now a major area of concentration for the introduction of new technologies. In the future, most office workers will use and be supported by computerized systems.

To prepare for the rapid changes that are taking place in offices, education and training of business students is changing rapidly. Increasing numbers of office workers are operating keyboards of computer terminals rather than typewriters. Instead of being required to shuffle mountains of papers, office

**Administrative jobs** will, increasingly, involve direct use of computers and computer terminals by managers and executives.

workers are dealing with electronically controlled information files. Computers will become vital tools in the offices of the future.

## PERSONAL SERVICES/LIFE-STYLE

You are already experiencing the effects of computers in almost everything you do. Computers are at work all around you. As you travel on the streets of many cities, computers control the traffic lights. In stores, computers assure speed and accuracy as you make your purchases. Computers deliver television programs into your home. Computers also control the games that many people play. When you buy a hamburger and a soft drink, the computer rings up the sale. A computer terminal even counts your change properly. Forms of technology have helped to heat or air-condition homes and workplaces. Even automobiles have advanced heating and air-conditioning systems.

The point is that technology is present in almost everything you do—almost everywhere you go. It is important, then, that you understand the role of technology in your life. Further, as you move toward the role of working citizen, your understanding and knowledge of technology can help to determine how successful you become. Technology will play an increasing role in all of your tomorrows.

## Summary

Technology not only meets existing needs and wants, it also creates new opportunities for careers. Modern technologies affect your life in many ways now, while you are still young. As you grow into adulthood, you need to consider the potential impact of technologies upon any career you select.

In this chapter, you are offered a peek into several fields of work. In each area, modern technologies have changed the way work is performed. In agriculture, computers help farmers plan how to use their land efficiently. Infrastructure involves many technologies, from planning to construction. Manufacturing is changing rapidly, from the assembly line to robotics. Fewer jobs will be available in the area of production in the future.

Materials handling involves many industries and technologies. As transportation methods change, so will the way goods are shipped and handled. The aerospace/military field is a leading developer and user of technology. Also, this area is one that has trained many people in the new areas of technology. Communication, both in transmission and information, has been another leader in technological change.

The banking industry has used computer technology to broaden services to the public. Education, too, will see major changes in the near future. Preparing students to take their places in a technological world is a major challenge to all educators today.

In short, modern technologies affect your life now and will continue to do so. In the coming years, you will be making decisions about what skills you wish to build, what career interests you have, and how you want to plan your life. You will need to consider technology when making these decisions.

## Using Technology

1. What type of career do you think you would enjoy? Write your answer in a short composition. Discuss what influenced your choice. What type of skills might you need to gain to be successful in the field? What technologies might you encounter?

2. Computer technology is an area that has gained wide acceptance. However, many people are afraid of computers. Some people imagine that computers are capable of controlling people. How do you feel about computers in your life? Do you know how to use a personal computer? Would you like to learn? Why or why not? Write your answers in a composition.

3. Throughout this book, the consequences, both good and bad, of technology are discussed. You have been asked to imagine what life would be like without technology. Now, imagine that you can peek into the future. What might life be like in the year 2050? Describe the basic life supports: food, shelter, and clothing. Describe the transportation and other systems that serve people. Try to apply the knowledge you have gained about trends in technology in your composition.

## Technology Terms

infrastructure
materials handling
warehouse
microwave
laser
fiber optics

# GLOSSARY

## A

**adaptable** Able to adjust to change and new demands.

**adobe** A heavy clay or soil found in dry, arid regions; used for building.

**adrenaline** A hormone that stimulates and excites the circulatory system.

**adze** An axe-like tool used to shape wood. It has an arching blade set at a right angle to the handle.

**aerobic exercise** Active movement of the body to increase the pulse rate and tone muscles.

**agribusiness** All of the activities involved in farming, processing, and distributing farm products, including the manufacture and distribution of farming equipment and supplies.

**agriculture** The combined technologies for growing food, timber, and other non-food products and delivering those products to market, including cultivating plants and domesticating animals.

**agritechnology** The complex system of technologies that produces needed agricultural products.

**alphabetic** Arrangement of a set of items in the order of the letters of the alphabet.

**alternating current (AC)** An electric current that reverses its direction of flow on a regular basis.

**amino acid** Organic chemicals that contain nitrogen and are necessary for life.

**application software** Programs that control the processing for individual jobs.

**applied science** The use of scientific principles to achieve specific results.

**applied technology** Applied science. The application of scientific principles to achieve specific results.

**arch** A portion of a round shape; often used as a structural support.

**architect** A person who designs structures.

**arithmetic** The processes for handling numeric data: addition, subtraction, multiplication, and division.

**arithmetic/logic unit (ALU)** Part of the central processing unit that uses arithmetic to process data.

**armature loop** A rotating wire coil that moves within a magnetic field causing electrons to flow through the wire.

**artery** A vessel that carries blood from the heart to the body parts.

**artificial intelligence** The programming of some computers to study and respond to their own stores of data and information.

**atom** The smallest part of an element that can exist alone.

**atomic bond** The force of attraction that holds atoms and molecules together.

**atrophy** The decrease in size and strength of a muscle or organ because of disuse, as happens after illness or injury.

**audio output** Computer output resembling human speech.

**automatic teller machine (ATM)** A computer-connected device for direct banking transactions by bank customers.

**axle** A shaft on which a wheel or pair of wheels revolves.

## B

**bacteria** Microscopic one-celled organisms. Some bacteria cause diseases.

**bar code** Vertical bars and blank spaces printed on labels of products for automatic entry of encoded data into computers.

**base** A solid surface that holds up or supports a structure.

**battery** A device that produces electrical energy through chemical reactions.

**beam** A long, heavy piece of wood, metal, or concrete used to construct the framework of a building.

**belt driven** The transfer of power from machine to machine with a system of belts and pulleys.

**binary** The internal (machine) language of a computer expressed in a series of 1 (one) and 0 (zero) values.

**binary digit** A piece of information represented by a 0 (zero) or 1 (one) value for computer processing.

**biofeedback** A health care technique by which a person learns to control physical conditions through intense mental concentration.

**bit** A device that is used to bore or grind wood or metal. Also, short for *b*inary dig*it*.

**blood pressure** A measurement of the force that the blood exerts as it moves through the arteries.

**body temperature** The degree of heat natural to a living body. Normal body temperature is 98.6 degrees Fahrenheit (37 degrees Celsius).

**boil** To change a liquid into a gas by adding heat or by reducing pressure.

**boiler** A tank designed to hold water and sustain pressure when heated.

**bolt** A threaded metal rod with a blunt tip that is used with a nut to fasten materials together.

**bracket** A triangular, or L-shaped support that fastens or holds materials at an angle.

**brake system** A group of related parts that allow the driver to slow or stop a vehicle.

**broadcast seeding** The planting of seeds by scattering them over a large surface area.

**brush** A device within an electric generator that makes contact with the armature loop and carries electric current away from the armature.

**budget** A plan for the spending of money.

**bulkhead** A partition that separates compartments and gives support.

## C

**calculus** A mathematical method for calculating and dealing with estimates and approximations.

**calorie** The unit for measuring heat generated from burning a given amount of fuel. The human body "burns" its fuel through chemical reactions.

**cantilever** A beam extending from a central point that is designed to support a structure such as a bridge.

**carbohydrates** Sugars and starches that are found in certain food products.

**carbon dioxide** A gas that results from the combustion of organic substances.

**central processing unit (CPU)** The part of a computer that performs actual calculations and operates the computer.

**cereal** A food made from grain.

**chad** The pieces of material created when holes are punched.

**chaff** The husks or pods of grains or grasses that remain after they are threshed.

**chain saw** A portable saw powered by a small motor.

**checkpoint** A pause in a project at which progress toward goals and completed work is evaluated.

**chuck** A device that holds a tool or a material in place so it can be rotated.

**circle** A perfectly round form in which every point on the outer edge is the same distance from the center.

**circulation** The movement of materials, particularly liquids, through a course that returns to the point at which it began.

**circulatory system** The body parts that carry blood to and from organs and tissues.

**classify** To determine the nature or identity of a data item. Also, to arrange in groups according to a system.

**claw hammer** A hammer with a claw-like head that is designed for pulling nails.

**clothing** Garments worn to protect the body and to maintain body temperature.

**coherent light** Light rays that are in parallel movement and of the same wavelength, as in a laser beam.

**cold storage** A refrigerated place for storing perishable goods.

**combine** A machine that reaps and threshes crops at the same time.

**combustion** The process of burning fuel to produce light or heat.

**comparison** To check values of items to determine if one is greater than, less than, or equal to the other.

**compass** A device that uses magnetism to determine direction.

**compiler** A program that translates written instructions understandable by people into a machine-readable language.

**compress** To reduce through pressure the volume or size of a gas, liquid, or solid.

**compression stroke** The first upward movement of the piston that compresses and thoroughly mixes fuel and air.

**computer** A group of related devices that process data according to instructions contained in programs.

**computer architects** Specialists in the design of computer equipment and system software.

**computer-assisted design (CAD)** Use of computers to design products.

**computer-assisted manufacturing (CAM)** Use of computers to control the manufacture of products.

**computer input operator** Person who presents data to a computer.

**computer system** Collection of processing devices and their programs that function in coordination.

**conductor** A material through which electricity passes easily.

**conservation** The careful management and protection of natural resources to prevent unnecessary destruction.

**construction** The technologies involved in the building of structures.

**contour plowing** To plow furrows by following the shape of the land to keep water from running off.

**contracting** Drawing together in size or shortening in length.

**control unit** Part of the central processor that coordinates the computer according to the instructions in a program.

**cooling system** A group of related parts that prevent overheating of an engine.

**cotton gin** A machine that separates cotton seeds from the fiber.

**craft** An occupation or trade requiring specialized skills.

**crop duster** An airplane or helicopter used to spray fields with pesticides or herbicides.

**crop rotation** A system of growing different crops on the same land in different years. Crops are rotated to prevent depletion of the soil.

**crude oil** Petroleum in its natural, unprocessed form.

**cryogenic** Relating to the production of very low temperatures.

**cultivation** The activities involved in caring for growing plants, including weed and insect control.

**cumulative** Information value that increases as new items are added to previous totals.

**cycle** A series of events that recur regularly over a given time.

**cylinder** A chamber within an engine where fuel and air are changed into force.

## D

**data** Raw facts; the "raw materials" that can be processed to produce information.

**data bank** An information storehouse used by a computer.

**data capture** The encoding of data in a form usable by a computer.

**data entry** The actual presentation of data to a computer for processing.

**data item** A single fact that has not been related to other facts.

**data media** Any material on which data can be recorded in computer-readable form.

**deduction** A process by which specific knowledge is gathered from studying generalized knowledge.

**delphi process** A problem-solving technique that considers the suggestions or ideas given by a number of people to make a decision.

**demand** A need or want for goods or services.

**destination** The end point of a course or journey.

**diagnosis** The process of identifying a disease by studying a person's physical condition.

**diagnostic technology** The methods and equipment used to identify health problems.

**die** A tool that can mold or press a desired shape; matched to a punch.

**diet** Total nourishment taken into the body as food or beverages.

**digestive system** The body parts that convert food into usable fuels.

**direct input** Capturing and entering data in a single function as part of the source transaction.

**disk drive** A device that holds, accesses, and records information on magnetically coated platters.

**displace** To replace one element of a production system with another.

**display** A temporary presentation of computer output, usually on a video screen.

**distillation** A refining process that separates impurities from a substance by alternately heating and cooling it.

**domesticate** A process that adapts animals for living in a close relationship with and to the advantage of humans.

**drilling** Cutting precise holes in materials with a high-speed rotary tool. Also, a seed-planting technique.

**drip irrigation** A method of supplying water to crops. This system can be compared to a faucet and valves and is used frequently to water individual plants.

**drive** To move an object with force.

**driven pulley** A pulley that is connected to another pulley but is not connected to a power source.

**driving pulley** A pulley that is connected directly to a power source.

## E

**economic system** All of the factors included in the monetary (money) and production systems of a given area.

**economy** All of the exchanges of money for goods and services that take place in a given area.

**edible** Eatable; plant or animal products that can be consumed as food.

**electrical current** The flow of electrons through a conductor.

**electrical system** A group of related parts that delivers energy necessary to produce power.

**electricity** A form of energy created by the movement of electrons.

**electrode** The source or destination of the flow of electricity in a battery.

**electrolyte** A chemical substance that has the ability to conduct the flow of electricity from one electrode to another.

**electromechanical accounting machine** An operator-controlled, nonelectric data processing device.

**electromechanical generator** A generator that produces electricity from mechanical power.

**electronic mail** The exchange of information transmitted between geographically separated computer systems.

**electronic publishing** The production of finished textual pages or corresponding microforms directly from computer files.

**element** A substance that cannot be separated into substances of other kinds.

**ellipse** A round shape that does not form a circle.

**endocrine** A chemical released in the blood stream that affects the functions of the body organs.

**endocrine system** The body parts that regulate the processes of the other body systems.

**energy** A source of power; the capacity for work.

**engine** A device that converts energy into mechanical power and motion.

**engineer** A person who plans and manages the building of a structure or device.

**environment** The natural surroundings of a specific area, including everything within, above, and below the ground.

**environmental protection** The efforts by people to safeguard the natural environment against pollution.

**ergonomics** Human engineering; a field concerned with designing equipment that can be used comfortably and efficiently by people.

**excretion** The elimination of waste from the body.

**excretory system** The body parts that gather and eliminate waste.

**executive keyboarding** Use of terminals with keyboards by executives for direct access to computer files.

**exhaust stroke** The upward movement of the piston that expels the burned gases from the cylinder.

**exhaustible** Limited, as in resource supplies that, in time, may be depleted, or used up.

**exploratory** An examination for diagnostic purposes.

**external combustion** The burning of fuel outside the cylinder where force is being produced.

## F

**factory** A building in which goods are manufactured, or produced.

**fastener** A device that joins materials to form a single structure.

**fat** A nutrient found mostly in oily or greasy foods. Fat contains many calories and is used by the body to supply energy and as a storage place for some vitamins.

**feedback** A return of corrected, updated, or evaluated information to its point of origin.

**fertile** In soil, the capability to sustain large quantities of plant life.

**fertilizer** A natural or man-made chemical that makes soil fertile.

**fiber** The threadlike, flexible strands that form parts of certain plant or animal cells.

**fiber optics** System or device that transmits images through a tube lined with light-conducting glass fibers.

**file** An organized collection of related information.

**fixed disk** Permanently mounted platters used to hold information for computer access.

**fixed pulley** A pulley that is attached to a block or holding device.

**flash frozen** To freeze foods rapidly in a cryogenic environment.

**flexible** The ability of a material to bend.

**flywheel** A wheel at the end of a crankshaft that transfers power from an engine to a belt or clutch.

**food** A substance that contains the nutrients the body needs to sustain life.

**food processing** The activities involved in receiving, preparing, and packaging crops for human use.

**force** Applied energy or effort.

**foundation** A prepared base that supports the structure of a building.

**four-stroke cycle** A series of events during which the piston moves four times—down, up, down, up—converting fuel to power.

**framework** A structure that supports and gives shape to a building.

**framing** The parts that support and give shape to a structure.

**free electron** An electron that is not attached to an atom and is able to move about freely.

**free-standing** A structure that supports itself.

**friction** A resistance to motion that results when a moving object is in contact with anything else.

**fuel** A material such as wood, coal, or gas that is burned to produce heat or power.

**fuel system** A group of related parts that store and supply fuel for an engine.

**fulcrum** The point of support at which a lever rotates.

**furrow** A long, narrow groove in the earth that is made by a plow or a wheel.

**fuselage** The central body of an airplane that holds the flight crew, the passengers, and the cargo.

## G

**gable** A triangular structure that is used to form a sloped, ridged roof.

**gangplank** A board or plank that forms a bridge by which passengers may board or leave a ship.

**gear** A toothed wheel that rotates on a central shaft and meshes with other gears to transmit power.

**genetic defect** A handicap or imperfection passed from a parent to a child.

**geothermal energy** Energy generated from the heat of the Earth's interior.

**grain** A plant whose seeds are used as a source of food for humans or animals.

**grain elevator** A tall, usually round structure or building used for storing grain.

**graphic output** Drawings or images created by a writing device that is controlled by a computer.

**gravity** A force of attraction exerted by a mass, such as the Earth.

**gravity flow** The tendency of matter to move from a higher elevation to a lower elevation. An example is the way water naturally flows downhill.

**grinding tool** A tool that is used to shape wood or metal by removing material through friction.

**ground water** A source of water that is found beneath the surface of the earth.

## H

**half-life** The time it takes for half of the skills or knowledge of a trained person to become obsolete.

**hammer** A tool with a solid head attached to a handle; used for pounding.

**hardware** The computer equipment in an information system.

**harvesting** The gathering of crops.

**health** The state or condition of a living body. The term usually implies the absence of disease or sickness.

**health care** The efforts involved in keeping people well and fit.

**health maintenance organization** An insurance group that provides preventive, diagnostic, and remedial health care for its members on an ongoing basis.

**herbicide** A substance used to kill plants, especially weeds.

**high-rise** A very tall building that has many stories.

**hinge** A two-part device that fastens materials together and allows them to swing, or rotate.

**hogan** A round adobe structure built by American Indians in the Southwest.

**holistic** Considering a condition or situation as a whole.

**holistic medicine** Treatment that involves the whole person—in mind, body, and spirit.

**hormone** A chemical released in the blood stream that affects the function of body organs.

**hub** The central part of a wheel.

**human resource** People functioning as part of a total life-support system.

**hydraulics** The study of force as derived from the control and movement of liquids.

**hydroelectric** Having to do with the production of electricity by water power.

**hypothesis** A proposed idea or assumption that attempts to state a principle of nature.

## I

**igniting** Starting combustion.

**immunity** Resistance to a medicine, or a disease.

**implement** To carry out or complete a plan.

**implementation** All of the steps taken to carry out a plan.

**impulse** A signal or message carried by the nervous system from one body part to another.

**inclined plane** A plane that forms an upward angle from a level surface.

**inductive** A process by which generalized knowledge is gathered from studying specific knowledge.

**inert** Lifeless; lacking the power to move.

**inexhaustible** Unlimited supplies, as with a resource such as air.

**infant mortality rate** A measurement of the number of deaths to babies less than one year of age for each 1,000 births.

**inference** An idea formed from a logical evaluation of existing evidence.

**inflexible** Unyielding; unable to adjust to change.

**information** Data that have been given meaning by processing.

**information processing cycle** The procedures that take place when data are converted to information. These include input, processing, output, and storage.

**information processing system** A coordinated set of data, procedures, people, and equipment that collects, processes, stores, and delivers information.

**information society** A society that depends on use of information and in which a large percentage of workers are in information occupations.

**infrastructure** The support systems, such as roads, buildings, and utilities, upon which an area depends.

**inoculate** To inject a vaccine into a body to help it build an immunity to disease.

**input** Entry of data into a computer system.

**intake stroke** The first downward movement of the piston that draws fuel and air into the cylinder.

**internal combustion** The burning of fuel within a cylinder in which energy is being released.

**internal combustion engine** An engine that produces energy by burning fuel within its cylinders.

**intrusive** To break in. In health care, this refers to practices that require "breaking" the skin, such as surgery.

**invention** The process of using the imagination to discover ways to provide for human needs and wants.

**inventory** Salable goods available for delivery to customers.

**irrigation** A way of supplying water to land that lacks sufficient rainfall to support crops.

**isometric exercise** Straining muscles in a planned pattern to build physical strength.

## J

**jet** System of propulsion (drive) based on emission of a stream of combusted gases.

**jig** A device that is used to keep a workpiece in the correct position for processing.

## K

**kernel** The grain or seed part of a cereal plant.

**kinetic energy** Energy that results from a body of mass in motion, as in a falling rock.

**knowledge** The body of facts, ideas, and principles people have acquired over time through processing and application of information.

## L

**laser** A high-powered, coherent beam of light.

**lathe** A machine that shapes or cuts wood or metal by turning it rapidly and pressing the workpiece against a tool.

**least cost feed blending** A computer-controlled process that gives animals maximum food value for the least cost.

**legislation** Acts of a law-making body.

**legume** A plant that stores its seeds in pods. Legumes often are grown to fertilize fields because of their ability to fix nitrogen in the soil.

**lever** A rigid object that rotates on a fulcrum. Pressure applied to one end increases the power delivered at the other end.

**leverage** The use of a lever to increase power.

**life expectancy** The number of years people can be expected to live, based upon statistics.

**life sign** An indicator of the condition of a living body.

**life span** The number of years people live.

**life support** A device or method that helps to sustain a living organism.

**load resistor** Any material or device that impedes (resists) the flow of electric current.

**lodestone** An ore that has magnetic properties.

**lubrication system** A group of related parts that store and supply oil to an engine to reduce friction.

## M

**magnet** A device or object that attracts ferrous materials.

**magnetic field** The space around a magnetic body in which magnetic attraction is present.

**magnetic pole** The end points, north or south, of a magnet.

**magnetism** The ability of certain devices or metals to attract ferrous materials.

**marketing** The activities involved in promoting and delivering goods to consumers.

**material** A substance that is used to produce goods.

**materials handling** Methods of storing, shipping, and delivering products.

**matter** The materials that make up the physical universe.

**mechanical device** A machine that uses or applies power to accomplish a purpose.

**mechanical energy** Energy produced by a machine or a tool.

**mechanics** The application of tools and machinery to do work.

**mechanism** A device that people use to produce power or energy.

**medium of exchange** An item with commonly accepted value that is exchanged for goods and services.

**methanol** A form of alcohol that can be a substitute for gasoline in internal combustion engines.

**microorganism** An organism that is invisible without the use of a microscope.

**microscopic** Visible only with a microscope.

**microwave** A short electromagnetic wave, used for transmission of signals.

**migratory labor** People who travel from place to place to find work, especially harvesting crops.

**milling** The process of cutting or grinding hard materials into a desired form.

**mineral** An inorganic substance that, in small doses, helps living things function properly.

**model** A representation of a process or a system through diagrams or equations.

**modeling** The creation of example situations by which many different courses of action can be tested and the best one chosen.

**modification** A change, usually for improvement.

**molecule** The smallest particle of an element or compound that still retains the characteristics of that element or compound.

**momentum** The force created by an object in motion.

**monetary system** An accepted group of procedures that set a value for and regulate standards for a medium of exchange.

**monoculture** A method of farming that devotes large areas of land to the cultivation of a single, specialized crop.

**movable pulley** An unattached pulley that is used to alter the direction in which a rope or belt moves.

**mulching** Spreading leaves, straw, or other materials around plants to prevent weed growth and water evaporation.

**muscular-skeletal system** The body parts that support and move a body.

## N

**natural resource** A needed material found in nature.

**necessities of life** The minimum requirements needed to sustain life: food, clothing, and shelter.

**negative electrical charge** An electrical condition in which there is a surplus of electrons. The symbol for a negative electrical charge is a minus (−) sign.

**nerve** A string-like band of tissue that connects body parts and passes impulses or messages from one part to another.

**nervous system** The body parts that receive stimuli and transmit impulses or messages from one body part to another.

**network** Combinations of circuits and wires that deliver electrical energy from generators to users.

**nitrogen** A gas that is a part of all living tissue and makes up 78 percent of the atmosphere.

**north pole** The positive pole of a magnet.

**nourishment** Food that provides essential nutrients that the body needs to function and grow.

**nucleus** The central part of an atom that carries a positive electrical charge.

**numeric** Related to numbers.

**nut** A device containing a hole with an internal screw thread that is used with a bolt to fasten materials together.

**nutrient** A substance in foods that provides living things with fuel and body building materials.

## O

**orbit** The path followed by one body that revolves around another.

**orchard** An area in which fruit or nut trees are cultivated.

**order of magnitude** A measure of quantity or quality that is 1,000 times greater than a given level.

**ordinance** A law enacted by a local legislative body.

**orient** To determine a position relative to a known object.

**outer shell** The shell of electrons in an atomic structure that is farthest from the nucleus.

**output** The production and delivery of information from a processing system.

**oxygen** A colorless, odorless, tasteless gas that is essential for combustion and as a life support.

## P

**parallel** A condition in which two lines are equal distances from one another along the entire length of the lines.

**paralysis** The loss of use or movement of a muscle or a part of a body.

**pentagon** A form that has five sides and five angles.

**perimeter supported** A structure in which the entire framework is located at the outer edges.

**permanent** Stable; lasting for a long time without moving or changing.

**perpendicular** A condition in which two lines cross to form 90 degree, or right, angles.

**pesticide** A chemical used to kill insects or animals that destroy crops.

**petroleum product** A good that is made from crude oil.

**pharmaceutical** A substance intended to overcome and prevent disease.

**phase** One of the steps in a series that make up a project.

**photovoltaic cell** A device that converts light into electrical energy.

**physical therapy** Careful, controlled exercise that helps bring a body back to health after an illness or injury.

**piston** A round metal part that moves in a cylinder to compress an air/fuel mixture or a gas.

**piston rod** A rod connected to a piston that tranfers force from the piston to another connecting rod.

**plan of work** A step-by-step outline of the tasks to be done to reach a goal.

**plane** A tool with a cutting edge protruding from a flat surface that is used to smooth, flatten, or shape wood.

**planting** The placing of seeds into prepared ground.

**plowing** Cutting and turning up soil.

**pneumatics** The study and application of gases under pressure.

**point of origin** The position at which a course or action begins.

**pollution** The contamination of the environment with waste produced by people.

**population density** The number of people who live within a given area.

**portable** Capable of being moved or carried.

**positive electrical charge** An electrical condition in which there is a deficiency of electrons. The symbol for a positive electrical charge is a plus (+) sign.

**potential** A condition that can—but has not yet—come into being.

**potential energy** Energy ready for use, such as a coiled spring or water behind a dam.

**power** The rate at which energy can be applied to do work.

**power cycle** A group of activities that change energy into force and then apply this force to do work.

**power plant** An engine or a set of related parts that applies energy to do work.

**power stroke** The second downward movement of the piston that results from the burning of the fuel/air mixture.

**power system** A complete set of people, procedures, equipment, and fuel that develops energy to perform work.

**preventive medicine** Treatment designed and used to avoid illness or disease.

**process** A series of steps followed to reach a known result.

**processing** The techniques that are used to convert data to information.

**producer** A person or organization that converts resources into marketable goods.

**productivity** A measurement of the rate of work output that can be delivered by people or machines.

**program** A set of coded instructions entered into a computer to direct processing and handling of data or information.

**progress** An improvement in people's ability to satisfy needs and wants.

**project** A series of steps followed to reach a known end result.

**project structure** The organization of steps of a project.

**protein** A nutrient made up of amino acids that provides the building blocks for cells in living things.

**proton** A fundamental particle of the nuclei of all atoms; carries a positive electrical charge.

**prune** To remove parts of a plant or a tree to increase crop production.

**psychological illness** A unhealthy mental condition.

**pulley** A small wheel with a grooved rim used to change the direction of, the amount of, and the application point of a lifting force.

**pulse** A measurement of the rate at which the heart is beating.

**punch** A device that makes holes or impressions in materials; matched to a die.

**punched card** A thin piece of cardboard that is punched with holes to encode data values.

**pure science** A systematic approach to gaining knowledge for its own sake.

**pyramid** A structure that consists of a square base and four triangular sides.

## R

**radiograph** A picture produced by X-ray equipment.

**radius** The distance from the center of a circle to the outer edge.

**rafter** A flat structural board that supports a roof.

**ramp** A sloping passage that joins different levels of a road or building.

**rate** A method of measuring something in terms of something else. For example, to determine the speed of an automobile, measure how many miles the automobile traveled in an hour.

**ratio** A relationship in quantity, size, or amount between two or more items.

**raw material** An unprocessed material that can be changed into a marketable good.

**ray** One of the lines of light that come from the sun or other source.

**reaper** A machine that cuts down a crop at harvest time.

**rectangle** A form that is made from four straight lines that are joined at 90 degree angles.

**recycle** To return processed materials to their raw form for reuse.

**refinery** A facility where crude oil is processed.

**regulate** To bring order; to bring under the control of law.

**renewable** Able to be restored, as in a natural resource that replenishes itself.

**reproductive system** The body parts involved in generating new life.

**resistance** An opposition; as in impeding the flow of electrons.

**resource** A source or means for meeting needs or wants; can include natural materials, energy supplies, or human thought or actions.

**respiration** The act of bringing oxygen into and releasing carbon dioxide from the body.

**respiratory rate** The number of breaths a person takes per minute.

**respiratory system** The body parts that carry oxygen to and carbon dioxide from the blood.

**rest energy** Energy that is contained in inert objects that lack the relative position of potential energy. An example would be coal under the ground.

**right angle** An angle formed when one straight line is perpendicular to another, forming a 90 degree angle.

**rim** The outer, circular part of a wheel.

**ring** A full shell of electrons orbiting within an atom.

**robot** A machine that operates by itself and performs a productive task.

**robotics** The use of equipment that operates itself and is controlled electronically.

**rocket engine** An engine that uses special fuels that create a thrust through rapid combustion.

**root crop** Plants with edible roots.

**roughage** Coarse food, high in fiber and low in nutrients, that helps the body's digestive and excretory systems.

**row crop** A type of crop that is grown in rows; such as cotton, lettuce, or grapes.

## S

**saw** A tool with a metal blade, the edge of which consists of a series of sharp teeth. A saw is used to cut hard materials such as wood, metal, or bone.

**scientific method** The process of stating a hypothesis, or idea, testing this hypothesis with an experiment, and then reporting the results of the experiment.

**scissors** A tool that cuts material by means of two opposing blades.

**screw** A threaded metal rod with a pointed tip that is used to fasten materials together.

**screwdriver** A tool that is used to tighten, loosen, or turn a screw.

**shell** A circular path in which groups of electrons revolve around a nucleus.

**shell structure** A structure in which the outer cover serves as the framework.

**shelter** A structure that protects people from danger and bad weather.

**skyscraper** A very tall building that has many stories.

**sledge hammer** A hammer with a large, heavy head; used with both hands.

**slip ring** A ring mounted on the shaft of a generator that holds the armature loop and permits it to spin.

**Socratic method** A way of asking organized questions in order to understand and build knowledge; named for its inventor, Socrates.

**solar energy** Energy that is produced by the sun.

**soldering** The joining of metal parts with a molten alloy of lead and tin.

**solution** The answer to or elimination of a problem.

**source of energy** A material or resource that has the potential to produce energy.

**south pole** The negative pole of a magnet.

**species** A group of plants or animals that have the same characteristics and distinguishing physical features and can be interbred.

**spire** A roof structure that forms a tall, pointed pinnacle.

**spray** A device that forces a jet of water through many small holes to create a rain-like mist.

**sprinkler system** A group of devices, including pipes and nozzles, used for irrigation.

**sprocket** A toothed wheel that engages with the holes in a chain or other material, as on a bicycle.

**square** A form that is made from four straight lines of equal length that are joined at 90 degree angles.

**stable** A condition that has remained unchanged for some time and may resist change.

**state** A condition that exists at a certain moment in time.

**stationary** To remain in the same place: immovable.

**steam engine** An external-combustion engine that uses steam to produce a force.

**steering system** A group of related parts that allow the driver to control a car's direction of movement.

**sterile** Pure; free from germs or other living microorganisms.

**stethoscope** An instrument used to listen to sounds within the body.

**stiffener** A rod that is used to give support.

**stimuli** Physical conditions or substances that cause responses in living beings.

**story** One floor or level of a building.

**straw** The hollow stems or stalks that remain after grain is threshed.

**stringer** A rod that is used to give support.

**stroke** The vertical travel of a piston in a cylinder.

**structural design** The planning that goes into building a structure.

**structure** A building that accommodates people or a casing for equipment.

**subatomic** A particle that is smaller than and is normally a part of an atom.

**substance abuse** The excessive use of drugs or alcohol.

**sugar** A sweet material that consists mostly of sucrose. Sugar is made from sugarcane and sugar beets.

**supplement** An addition to an amount of a substance. For instance, vitamin pills are taken as an addition to the vitamins received by eating food.

**surplus** The amount of a good or service that exceeds existing demand.

**suspension** To be hung or supported from above.

**suspension system** A group of related parts that allow a car to absorb the shock of bumps in a road.

**system** A group of people, procedures, equipment, and resources that function together to produce desired results.

**systems approach** An application of systems, or systematic methods.

**systems development** A structured method, used to create new systems.

**systems of technology** A group of technologies used in combination to deliver products or services that meet demands.

## T

**tap** A device that is used to form an internal screw thread in a hole.

**technology** The application of scientific knowledge for a specific purpose.

**teeth** Parts that jut out from a gear that can fit into other gears so that sets of gears work together.

**temperature** Degree of hotness or coldness.

**terraces** Flat platforms of earth with sloping sides that rise one above the other on a hillside or mountain.

**terracing** To make flat, level platforms of earth on sloping hillsides or mountains.

**textile** Cloth created from thread, yarn, or fiber.

**therapist** A person trained to help people recover from illness or injury.

**therapy** A remedy for an illness or a way to restore health after an illness.

**thermometer** An instrument used to measure temperature.

**threshing** The act of separating grain or seed from the plant.

**thrust** To drive or push with force.

**tilling** Plowing land to prepare for crop planting and growth.

**tool** A device that assists a person in the performance of work.

**torque** A force that produces rotation.

**transferable skills** Knowledge or training that a person can use in more than one job.

**transportation** The process or means of travel from one place to another.

**trapezoid** A four-sided form in which only two sides are parallel.

**trauma** The result of physical injury to a living body.

**treatment** A method used in dealing with a health problem.

**triangle** A form that is made from three straight lines and three angles.

**triangular shape** A shape that has three straight sides and three angles.

**truck garden** A garden of row crops, such as strawberries and many vegetables. The term refers to an older, traditional way of farming—when farmers would bring crops into cities on trucks.

**turbine** A device that produces power through a rotary motion in response to the passage of gases or liquids.

**typesetting** The process by which letters, numbers, and symbols are represented in specific sets or fonts.

## U

**unskilled** A lack of qualifications for employment.

**urbanization** The movement to and settlement of people in highly populated areas.

**utopia** An imaginary place with ideal laws, government, and social conditions.

## V

**vaccine** A substance that helps a body resist or build immunity to a disease.

**vehicle** A device used for transporting people or things.

**vein** A vessel that carries blood from the body parts back to the heart.

**vessel** A hollow, tube-like canal that holds and circulates fluid such as blood.

**vines** Plants with long, slender stems that grow along the ground. The stems of these plants cannot support the weight of the plant.

**vital sign** An indicator of the condition of a living body: the pulse rate, temperature, respiration rate, and blood pressure.

**vitamins** Nutrients that a human body needs to absorb food and to put other elements to work.

## W

**water table** The supply of ground water in an area.

**welding** The joining of metal parts by melting the edges of the parts and allowing the molten metal to flow together and harden.

**wheel** A circular frame or solid disk that is capable of turning on a central axis.

**work** The result of using force to move an object or complete a task.

**workpiece** Units of material upon which machining or other production operations are performed.

## X

**x-ray** Wave of electromagnetic energy that passes through solid matter and records an image on film, making possible nonintrusive examination of the human body.

## Z

**zoning law (ordinance)** A law that regulates use of land.

**zoning ordinance** See zoning law.

# INDEX


AC. *See* Alternating current
Accidents, 218, 221
Adaptability to change, 311
Administration, 382–383
Adobe, 137–138
Adrenalin, 214
Adze, defined, 78
Aerobic exercise, 244–246, 247, 250
Aerospace industry, 376–378, 384
Aging, and health, 218, 232
Agribusiness, defined, 162
Agriculture: computers in , 162–163; crops (*see* Crops); defined, 157; development of, 12–13; technology of (*see* Agritechnology)
Agritechnology, 183–204, 368–370, 384; computers in, 369; defined, 157, 180, 203; developments in, 12, 26; materials used in, 58–59; system components, 160, 180
Aircraft: building of, 377; for crop dusting, 190; developments, 38; invention of, 9; lighter-than-air, 37; power sources, 107–109; technological systems in, 338; World War II, 266
Air pollution, 105, 111, 222
Alcohol production, 168
Alfalfa, 169–170, 180, 188
Alphabet, development of, 42
Alphabetic organization, 271
Alternating current (AC), 126, 129
ALU. *See* Arithmetic/logic unit
Animals. *See* Livestock
Application programs, 283–284
Applied science, 335–336, 369
Applied technology. *See* Applied science; Technology
Arch: defined, 137; Roman, 138
Architect, work of, 41
Architecture, 371
Aristotle, 18
Arithmetic functions, in data processing, 257, 270, 296
Arithmetic/logic unit (ALU), 296, 302
Armature loop, 125, 129
Armed Forces, 376–378
Arteries, 209, 214
Artificial intelligence, 328–329, 330; defined, 328
Assembly line, 344–345
Athletes, 246; caloric intake, 241; computer studies of, 328
ATM. *See* Automatic teller machines
Atomic structure: copper, 121; hydrogen, 119; and magnetism, 122–123
Atoms, 119; outer shell, 121; rings, 121; shells, 121; stability of, 121
Atrophy, defined, 248
Audio output, 298–299
Automatic teller machines (ATM), 288, 295, 380–381
Automobiles: computing devices in, 318; development of, 10–11; environmental problems caused by, 222–223; robotic manufacture of, 324; systems in, 105–107, 336–337
Avocados, irrigation of, 193
Axe, 77
Axle, 85

Bacon, Francis, 19, 20
Bacteria, in milk, 176
Balloons, 37
Banking industry/banks, 288, 295, 380–381, 384
Bar code, 259
BASIC, 284
Batch output, 299
Battery, 121–122
Beams, 141, 143
Beans, 161, 169
Bell, Alexander Graham, 21
Belt-driven machines, 98
Benefits of technology, in relation to costs, 306–307, 330, 352, 353–354
Berries, 171
Binary digit. *See* Bit
Binary language, 284
Biofeedback, 238
Biotechnology, 155–251; systems approach to, 348. *See also* Health care; Health maintenance
Bit, defined, 291
Blood chemistry tests, 229
Blood pressure, 214, 231, 237
Blood vessels, 209
Body systems, human, 209–214
Body temperature, defined, 12
Boiler, 96
Bolts, 82
Brackets, 82–83
Brain, 210–211, 248
Brake system, 107
Breeding and development, 160, 201–203, 204; records, 175
Bridges, 147
Broadcast seeding, 188
Brushes, generator, 125, 129
Budget, defined, 354
Buildings. *See* Structure(s)
Bulkhead, 150
Business, computers in, 320–323
Business education, 312–313

CAD. *See* Computer-assisted design
Calculators, 258, 273
Calculus, 20
Calories, 240, 241
CAM. *See* Computer-assisted manufacturing

Canals, 33, 36
Cancer, 220–221, 223; x-rays and, 229–230
Cantilever construction, 147
Carbohydrates, 240, 250
Carbon dioxide, 210
Career selection, 350–351
Carrots, 168
Cassavas, 161
Cassettes, magnetic, 293–294
Cattle, 173–174, 176. *See also* Breeding and development
Caves, 131–132, 151
Cells, 121
Census. *See* United States Bureau of Census
Central processing unit, 295–296
Cereals, 161, 180. *See also* Grain
Chad, defined, 80
Chaff, defined, 196
Chain saw, 179
Change(s): adaptability to, 311; in job skills, 308–309; technology and, 9, 263, 304–330
Character recognition, 294, 329
Checkpoint, defined, 354
Chemical diagnosis, 228–229, 235, 237
Chrysler Corporation, 353
Chuck, 80
Circle, defined, 137
Circular structures, 137–138
Circulation, defined, 209
Circulatory system, 209, 231; and exercise, 244
Cities, growth of. *See* Urbanization
Classification, 277
Cloth, production of, 30–31, 33
Clothing: and agritechnology, 157; need for, 12–13, 22, 27–28
Coal, 63, 117
COBOL, 284
Coconuts, 161, 170–171
Coherent light, 185
Cold storage, 198
Colds, 220
Columbus, Christopher, 17
Combine, development of, 196
Combustion, 95; basic requirements for, 100–101, 103; and body function, 210; defined, 62; principles of, 100–101, 110; temperature of, 104
Combustion systems, 95–96; external, 96–99, 110; internal, 99–104, 110
Comfort(s), desire for, 15–16, 22, 50
Commerce, 380–381
Communication, 384; written, 42–43, 45–46
Communication industry, 378
Communication signals, 378
Communication technology, development of, 42–43, 45, 47
Comparison function, in data processing, 270–271, 296
Compass, defined, 124
Compression, in engine operating cycles, 102, 107–108
Compressors, 107–108
Computer(s): in agribusiness, 162–163; in agritechnology, 369; applications of, 317–329; change and, 313–315; costs of, 267, 285–286, 317; defined, 258, 277, 285–286; early developments, 45; educational uses of, 381–382; first practical use of, 45; generations of, 315, 317; hardware, 383–386; in homes, 383; information explosion and, 263–269, 274; in livestock management, 175; manufacture of, 337–338; as tools, 262–263, 268
Computer architects, 314, 337–338
Computer-assisted design (CAD), 325–326, 330, 338
Computer-assisted manufacturing (CAM), 325–326, 330
Computer input operators, 278
Computer models, 339, 348; in agriculture, 162–163
Computer operations, 286–301; defined, 278. *See also* Input; Output; Processing; Storage
Computer specialists, 314, 337–338
Computer systems, 277–302; defined, 277. *See also* Computer(s)
Conductor, electrical, 120
Conferences, electronic, 323
Conservation of resources, 65
Construction, 41, 370–371
Contour plowing, 187
Contraction, defined, 212
Control(s), 54–57; over disease, 220; of distribution, 54–55; governmental, 306; need for, 11–12; of quality, 342–343; in systems, 340; of use of products, 54–55. *See also* Monitoring
Control unit, of computer, 296
Conveyors, 198
Cooling system, 106
Copiers, xerographic, 45
Copper atoms, 120–121
Corn, 163–164; breeding and development of, 202
Costs of technology: in relation to benefits, 306–307, 330, 352, 353–354; of computer power, 267, 285–286, 317; defined, 51
Cotton gin, 178
Cotton industry, 177–178
Cottonseed oil, 177
Crafts: defined, 26; located near resources, 28–29; specialization in, 26–27
Creativity, 7
Crop dusters, 190
Crops: basic groups, 161–162, 180; fuel from, 168; protection of, 165; rotation of, 164–165, 169, 180
Crude oil, 50
Cryogenic methods, 199
Cultivation, 160, 180, 203; defined, 188
Cultivators, 189
Cumulative, defined, 261
Cutting tools, 77–79, 90; development of, 77–78; sharpening of, 80
Cycles, of internal combustion engine, 101–103
Cylinders, 36, 99, 103; defined, 96; diesel engine, 104

Daedalus and Icarus, 7–8, 10
Dairy industry, 173–174, 176
Dairy products, storage of, 198
Dams, 116, 117, 191
Data: comparisons, 257; computers and, 256–257; defined, 256, 279; entry (*see* Data entry); files of, 271; importance of, 280; transcription of, 286–287, 288; transposition of, 270
Data banks, defined, 319
Data capture, 286–287, 295, 302; defined, 289; equipment, 289–294
Data entry, 295; defined, 287; methods, 287–288. *See also* Input
Data items, 256–257, 273
Data processing. *See* Processing
Data processors, people as, 268
Decision making, 351–353
Deductive reasoning, defined, 19
Delphi process: origin of, 18; use of, 18–19
Delphic Oracle, 18–19
Demand, 50–51, 67; changes in, 50; defined, 9, 50; for gold, 55
Design: general, 357–358, 364; technical, 359–360, 364
Destination, defined, 33
Diagnostic techniques/technology, 227–228, 232; computers applied to, 230; in health maintenance, 235; improvements in, 229–230

Die (tool), 80
Diesel engine(s), 103–104, 110; fuels, 103
Diet, 236, 238–243, 250; defined, 238
Digestive system, 212, 231, 243
Direct input, defined, 295
Diseases, 218–221, 232; of childhood, 237; control of, 39, 46
Disk: data recording patterns on, 292; as magnetic media, 291–292; types of, 292
Disk drive, 292
Displacement of workers, 308–311
Display, video, 298
Distilling techniques, 168
Documents, 297–298, 302
Domestic, defined, 12
Drilling, 80; techniques, in planting, 188; tools, 80, 90
Drip irrigation system, 192–193, 204

Economic system, defined, 51
Economy, defined, 51
Edible plants: basic crops, 161–162, 180; defined, 161
Edison, Thomas, 20–21
Education, computers in, 381–382, 384
Eiffel Tower, 41
Electrical charges: negative, 120; positive, 119
Electrical current, defined, 121, 129
Electrical systems, 105
Electricity, 119–128, 306; as source of energy, 32, 46, 109; sources of, 119, 124–125
Electrodes, 121–122
Electrolytes, 122
Electromechanical: accounting machines, 264; generators, 122
Electronic conferencing, 323
Electronic engineers, 314
Electronic mail, 321–322
Electronic publishing, 322–323
Electronics, 372; advances in, 266–267
Electrons, 120; flow of, 121; subatomic size, 120
Elevators, 42, 145
Ellipse, defined, 138
Encoded documents, 294, 302
Endocrine(s): defined, 213; system, 213–214
Energy: application of, 93, 110; defined, 30, 114; potential, 30; in relation to power, 118–119, 126.
Energy sources, 30–32, 46, 49, 59, 62–64, 95, 110; defined, 57; natural, 113–115, 118, 128; for power systems, 109; and processing of materials, 57
Energy states, 115–117, 128; changes in, 117
Engineer, work of, 41
Engineering, human (ergonomics), 326–328, 330; mechanical, 372
Engines: defined, 36; external combustion, 36; internal combustion, 36–37 (*see also* Automobiles); invention of, 52
Environment, defined, 56
Environmental problems, 222–224, 232
Environmental protection, 56
Epidemics, 220
Ergonomics, 326–328, 330
Evaluation, 363; of feasibility, 347, 351–354, 364; of project, 347, 362–363; of strengths and weaknesses, 356
Evaporation, 103
Exception, defined, 342
Excretion, defined, 212
Excretory system, 212–213, 228, 231
Executive keyboarding, defined, 321
Exercise, 236, 243–247, 250; for health maintenance, 248; therapeutic, 247–248
Exhaust stroke of gasoline engine, 103
Expectations, 356, 357
Exploratory surgical operations, 229
Exterior walls, 144
External combustion, defined, 96
External combustion engines, 96–99, 110; defined, 36; steam engine, 96–99

Factories: defined, 27; energy and, 31–32; location of, 28–29, 33; steam engines in, 98–99; structural design of, 146. *See also* Manufacturing
Factory workers, needs of, 29
Farming. *See* Agriculture; Agritechnology
Fasteners, defined, 81
Fats, 241–242, 250
Feasibility evaluations, 347, 351–354, 364
Feedback, 341–343, 363, 365; defined, 341–342; and monitoring, 341–343
Fertile, defined, 186–187
Fertilizer, 186
Fiber, in diet, 242
Fiber optics, 378
Files: defined, 271; storage of, 271–273
Fire, 131
Fish farming, 175
Fixed disk, 292
Flash freezing, 199
Flexible materials, 137
Flight, 7–9. *See also* Aircraft; Space travel
Flywheel, 98
Food: agritechnology and, 157; costs, 194; foreign-grown, 195; fuel value, 239–240; need for, 12–13, 25, 27–29; technology of production, 29–30, 59
Food processing, 76, 160, 198–201; cottonseed oil in, 177; defined, 198, 204; frozen food industry, 198–199
Food supply, 131–132; moving about for, 132–133, 137
Force, 93–94, 95–96; defined, 30, 93, 110
Ford, Henry, 371
Forestry, 178–179
Form, in relation to function, 151, 152
FORTRAN, 284
Foundation, structural, 133
Framework, structural, 137, 141, 143; defined, 134, 152; of perimeter-supported buildings, 146; of skyscrapers, 146. *See* Structures
Free electrons, 121
Friction, 85
Frozen food industry, 198–199
Fruit trees, 170
Fuel(s), 62–64, 100; alternative sources of, 158, 167–168; diesel engine, 103; food as, 210, 239–240; in power systems, 95, 110; in structures, 40; usage of, 104–105
Fuel system, 106
Fulcrum, defined, 83–84
Function, in relation to form, 151, 152
Furrows, 186; distances between, 187
Fuselage, 150

Gables, 136
Gangplanks, 86
Gases: compression of, 107–108; expansion of, 95, 99; inert, 169; polluting, 222–223
Gears: defined, 89; principles of, 89; ratios, 89; teeth, 89
General design, 357–358, 364
General Electric, 45, 301
Generations of computers, 315, 317

Generators, 125–126; electromechanical, 122
Genes, 225
Genetic, defined, 225
Genetic defects, 225
Genetic traits, 225, 232
Genetics, defined, 202
Geothermal energy, 62
Germs, 219, 220
Gin (cotton), use of term, 178
Goal setting, 347, 348, 350, 351
Goat raising, 175
Gold, price of, 55
Government records, 264. *See also* Social Security; United States Bureau of Census; World War II
Government regulation, 55–56, 306
Grain: farming of, 163–167, 180; grinding of, 29. *See also* Cereals
Grain elevators, 197–198
Grapes: picking, 171; processing, 172
Graphic output, 298, 299, 302
Gravity, 90; defined, 74–75; discovery of laws of, 20
Gravity flow irrigation system, 191, 204
Gray, Elisha, 21
Greek: ideas, 18; mythology, 7–8
Grinding: principles of, 79–80; tools, 79–80
Ground water, 191
Gutenberg, Johann, 42–43, 315

Hammer: claw, 76; mechanical principles of, 76–77; sledge, 76
Handling of food crops. *See* Storage of foods and goods
Hardware: computer, 284–286; defined, 283
Harvesting, 160, 165–166, 180, 195–196, 204; defined, 193; technological advances in, 195–196; tree crops, 170–171
Health and urbanization, 38–39
Health care, 207–232; defined, 207–208, 218; life expectancy and, 208; measures of quality of, 208, 217; of migratory laborers, 194; needs for, 13–15, 22; objective of, 207–208; scientific discoveries in, 39; technologies, 226–231, 232
Health insurance, 236
Health maintenance, 235–250; checkups, 236–237; defined, 235; organizations, 236, 249, 250; personal, 238–248; technologies, 235–238, 249
Heart rate, 244–246, 250
Heat, in power systems, 95
Herbicides, 189
Hexagonal structures, 142
High-rise buildings. *See* Skyscrapers
Hinges, 83
Hoe/hoeing, 188
Hogan, defined, 138
Hogs, 174
Holistic, defined, 238
Holistic medicine, 238, 249
Hollerith, Herman, 267
Home(s): computers in, 317–319; permanent, 133, 137; portable, 133, 137, 151; working at, 32. *See also* Structure(s)
Homicide, defined, 221
Hormones, defined, 213–214
Hospitals, 230
Hub, 85
Human body, 209–214
Human engineering (ergonomics), 326–328, 330
Human resources: defined, 49; in relation to technology, 50–57, 67
Hydraulics, 336
Hydroelectric plants, 117
Hydrogen, 119; as fuel, 64
Hypothesis, defined, 20

Icarus and Daedalus, 7–8, 10
Ignition, defined, 100
Illness, 207–208
Immunity, defined, 220
Impact printer, 298
Implementation, 52, 67, 347, 365; defined, 52, 361
Impulse, defined, 211
Inclined plane, principles of, 86
Inductive reasoning, defined, 20
Industrial Revolution, 371; steam engine and, 98–99
Inert gas, defined, 169
Inexhaustible resources, 65
Infant mortality rate, 217, 220
Inflexibility, and change, 311
Information, 257–274; analysis and use, 257–258; communication of, 378–379; cumulative, 261; defined, 257; importance to business and government, 255–256; sources of, 256–258
Information explosion, 20–21, 230, 263–268. *See also* Knowledge: accumulation of
Information processing cycle, 268–273, 274. *See also* Processing
Information processing systems: defined, 273; people in, 278–279; requirements for, 278, 286
Information services, computer access to, 219
Information society, 256
Information technology, 17–21, 22, 253–330
Infrastructure, 370–371, 384; defined, 40
Injector, 104
Inoculation programs, 238
Input, 302, 341; defined, 268–269, 274; forms, 281; procedures, 280–281, 286–288. *See also* Data entry
Inquiry, process of, 18
Insect control, 165, 188, 190, 204
Insurance: health, 236; programs, 250
Intake stroke of gasoline engine, 102
Internal combustion engines, 118–119; defined, 36; four-stroke cycle, 101–103, 110; principles of, 99–104; problems associated with, 104–105, 111
Intrusive diagnosis, 229
Invention, 52, 67; defined, 52; and technological change, 21, 22, 52
Inventory management: artificial intelligence applied to, 328; and robotic control, 324–325
Inventory reports, 260–261
Irrigation, 160, 178, 180; methods, 191–193, 204; need for, 190–191
Isometric exercise, 246–247

Jet engines, 107–108, 111
Jig, defined, 80
Job instructions, 282, 301
Jobs: creation of, 310–311; elimination of, 308–311

Keyboard entry, 287–290
Keyboarding, executive, 321
Keyboards, 302; separate, 327; of video display terminal, 298
Keystroking, defined, 289–290
Kidneys, artificial, 230
Kinetic energy, 115, 116, 117
Knife, 78
Knowledge: accumulation of, 17–21, 22, 42, 230, 257; and adaptation, 311–312; communication of, 42–43, 47; computer access to, 319; defined, 257–258; as foundation of technology, 17, 22. *See also* Information explosion

Land preparation, 160, 180, 183–187, 203; clearing, 184; planting in patterns, 187; surveying, 185; terracing, 185–186
Land use: for farming, 183; regulation of, 55–56

Laser, 378; defined, 185
Lathe, 80
Least cost feed blending, 163
Legislation, defined, 55
Legislative controls, 55–56
Legumes, 161, 169–170
Leonardo da Vinci. *See* Vinci, Leonardo da
Lever: defined, 76; principles of, 83–85
Leverage, 90; application of, 77; defined, 76
Life cycle, 213, 217–218
Life expectancy: defined, 13; health care and, 208
Life span. *See* Life expectancy
Life support, 12–13
Life-style, 355–356, 383
Life (vital) signs, 214–217, 230, 231, 237. *See also* Diagnostic techniques/technology
Linotype, 43, 45
Livestock, 12–13, 173–176, 180–181; breeding and development, 160, 175, 201–203, 204; feeding, 163–164; products, 200–201
Load resistor, 126
Local networks, 320–321, 323
Lodestone, 122
Looms, water power for, 31–32
Lubrication system, 106
Luxuries. *See* Comfort(s)

Machine language, 284
Magnetic cassettes, 293–294
Magnetic disk media, 291–293, 300–301; fixed, 292; recording patterns, 292; removable, 292–293
Magnetic diskettes, 293
Magnetic field, 124; defined, 123
Magnetic force, and electrical power, 124–125
Magnetic ink character recognition (MICR), 294
Magnetic poles, 124, 125
Magnetic tape, 291, 300–301; recording pattern, 291
Magnetism, 122–125
Magnets, 122–125, 129
Mail, electronic, 321–322
Main memory, 296–297
Management techniques, 18–19
Manuals, 282, 322
Manufacturing, 371–372, 384; computers in, 324–326. *See also* Computer-assisted manufacturing (CAM); Factories
Marketing, defined, 54
Materials, defined, 57, 68. *See also* Raw materials
Materials handling, 373–375
Mathematics, 20
Matter, 128; defined, 119
McCormick, Cyrus, 196
Meat industry, 200–201, 204
Meat processing, 200–201
Mechanical devices. *See* Mechanisms
Mechanical energy, defined, 109, 111
Mechanical engineering, 371
Mechanical principles: application of, 90; of hammer, 76–77
Mechanics, defined, 74
Mechanisms, 73, 90–91; defined, 74, 83, 90; examples, 74
Medical facilities, 230–231, 232
Medical technology, 14–15. *See also* Biotechnology; Health *headings*
Medium of exchange, defined, 51
Memory, 302; on microchips, 317, 330; versus storage, 300–301. *See also* Main memory
Mergenthaler, Ottmar, 43
Metal: tools for cutting, 77–79
Methanol, 158, 167
MICR. *See* Magnetic ink character recognition
Microchips, 317, 330
Microcomputers, 267, 317–319
Microfilm, 299
Microorganisms, 219–220
Microscopic, defined, 219
Microwave(s): defined, 266; transmission technology, 266, 378
Migratory labor, 169, 193–194
Military fields, 376–378, 384
Milk, radiation treatment of, 176, 200
Milking machines, 176
Milling, 29, 80, 198; tools for, 80, 90
Minerals, 242, 250
Missiles, 108
Modeling techniques, 162
Models, 348; in agriculture, 162–163; defined, 339
Momentum, 74, 76
Monetary system, 51
Monitoring, 339–343, 354–355, 365; and feedback, 341–343. *See also* Control(s)
Monoculture system, 164–167, 201; defined, 164, 180
Mortality rate, of infants, 217, 220
Movable pulley, 88
Movable type, invention of, 43
Movement, 93–95, 96–97
Movies, development of, 16
Mulching, 189, 204
Muscular atrophy, 248
Muscular contraction, 212
Muscular-skeletal system, 212

Natural energy, undependability of, 118
Natural resources: critical, 104, 306; defined, 49; and limitations on technology, 57. *See also* Energy sources; Materials; Resources
Necessities of life, defined, 12, 22. *See also* Needs
Needs: agritechnology and, 157–158; basic, 157; defined, 50; technology applied to, 9–21
Nerves, function of, 210
Nervous system, 210–211
Networks, 320–321, 323
Newton, Isaac, 20
Nitrogen, 169, 212–213
North pole, 124
Nourishment, defined, 239
Nuclear energy, 37, 62, 223–224
Nuclear radiation: for food sterilization, 176, 199–200; hazards, 63, 224
Nuclear reactors, 223–224
Nuclear waste disposal, 63, 224
Nucleus, 119
Numeric organization, 272
Nuts (food), 170–171; shelling, 171
Nuts (metal), 82; defined, 82

Ocean tides, 62
OCR. *See* Optical character recognition
Octagonal structures, 142
Office automation, 320–323, 382
Off-line, defined, 299
Onions, 168
On-line, defined, 299
Optical character recognition (OCR), 294
Orbit, 120
Orchards, 170
Orders of magnitude, 315, 317
Ordinances, 40, 47, 56
Orientation, defined, 363
Otis, Elisha, 42
Output, 271, 297–299, 302, 341; defined, 271, 274
Oxygen, 210; in combustion, 100–101

Paper, 157–158, 179; invention of, 42
Paper records, replacement of, 320, 330
Parallel, defined, 141

Paralysis, defined, 248
Parasites, 219
Parity bit, 291
Pascal (computer lanugage), 284
Pasteur, Louis, 176
Pasteurization process, 176
Payroll deductions, records required for, 263–265
Peanuts, 161, 169
Pentagonal structures, 142
Perimeter, defined, 146
Perimeter-supported buildings, 146
Perpendicular, defined, 141
Petroleum/petroleum products, 117; in relation to agricultural products, 167; alternatives, 158, 167; defined, 50; and demand, 50; limited supply of, 104; systems to produce and deliver, 52; use in internal combustion engines, 100
Pharmaceutical technology, 226
Photovoltaic cells: conversion of sun's energy by, 59, 62; defined, 114
Physical: checkups, 250; diagnosis, 229–230, 232; examinations, 236
Physical therapy, 247–248
Pig-farming, 174
Piston rod, 96
Pistons, 96–97; stroke of, 97
PL/1, 284
Plan of work, 354–357
Plane (tool): defined, 78
Planning, 347, 348
Planting: defined, 188; methods of, 188
Plants: breeding and development, 201–203, 204; diseases of, 188; edible crops, 161–162, 180; petroleum in relation to, 167
Plato, 18
Plow, defined, 186
Plowing, contour, 187
Pneumatics, 336
Point of origin, defined, 33
Pollution, 105, 111, 222, 306
Population, 40–41, 220
Potatoes, 161, 168
Potential energy, 30, 115, 116
Poultry farming, 175–176, 201
Power, 93–111; defined, 30, 93, 110; in relation to energy, 118–119, 126
Power cycle, 95
Power plant, 105
Power stroke of gasoline engine, 102–103
Power systems, 95–96, 110; defined, 95, 110; energy sources, 109; failure of, 126; and related systems, 105–107
Preventive medicine, 237–238, 249, 250
Primary storage. *See* Main memory
Printed output, 297–298
Printers, 297–298
Printing press, 43, 45, 315
Processing, 270–271, 295–297, 302, 341; defined, 257, 270; supermarket uses of, 259–261; types of, 270–271
Processor, 302; defined, 295
Producers, 29
Production, technology of, 27–32, 46. *See also* Manufacturing
Productivity, 307
Programming languages, 284
Programs, 282–284; data processing and, 258–259; defined, 277
Progress, technological, 6; effects of, 6. *See also* Change(s)
Project management, 347–348, 354–355
Projects: defined, 346; personal, 350–363; phases of, 347; structure of, 347–348
Proteins, 250
Protons, 119, 121
Pruning, 170
Psychological illness, 225
Psychological problems, 224–225
Psychology, defined, 225
Publishing, electronic, 322–323
Pulley: driven, 88; driving, 88; fixed, 87–88; movable, 88; principles of, 87–88
Pulse, 214–215, 231
Pulse rate, and aerobic exercise, 244–246
Punch, 80
Punched cards, 264, 290
Pure science, 335, 364
Pyramids, 134–135, 136, 137

Quality control, 279, 342–343

Radar, 266, 338
Radiation: in sterilization of foods, 176, 199–200; hazards, 63, 224
Radio communications, 266, 378–379; development of, 16
Radiographs, 229
Radius, defined, 137
Rafters, 136; defined, 136
Railroad locomotives, steam-driven, 96–98, 100
Railroads, 196, 306, 373
Ramp, principles of, 86
Ratio, defined, 89
Raw materials, 58–59, 68; defined, 30
Rays, of sun, 114
Reaper, invention of, 196
Reaping, defined, 196
Records: defined, 272; storage of, 271–273; transaction, 268
Rectangle, defined, 141
Rectangular structures, 140–141
Recycling, 66
Regulation, governmental, 55–56, 306
Removable disks, 292–293
Renewable resources, 64, 167
Reports: inventory, 260; sales, 260
Reproduction, 213
Reproductive system, 213
Reservation systems, 338
Resistance, 126
Resources, 48–68; critical, 104, 306; importance of, 49; limited supply of, 45–46; renewable, 64, 167; shortages of, 306; supplies of, 64–66; use of, 45–46. *See also* Human resources; Natural resources
Respiration, defined, 210
Respiratory rate, 216
Respiratory system, 210; exercise and, 244
Rest energy, 116–117
Retrieval of data, defined, 292
Rice, 163
Right angle, 140–141
Rim (of wheel), 85
Robot, defined, 324
Robotics, 324–325, 330, 351–352, 372, 384
Rocket engine, 107, 108–109, 111
Roof structures, 135–136; gabled, 136; rafters, 136; support for, 143
Root crops, 161, 168–169
Roughage, 242
Row crops, 172–173
RPG, 284

Safety: ordinances, 40; in relation to technological development, 10–12, 22
Sales reports, 260–261
Saw, 77–78
Schools, computers in, 381–382
Science, in relation to technology, 335–336, 369

Scientific method, 19–20
Scissors, 79
Screw, defined, 81
Screwdrivers, 81–82; Phillips head, 82; straight-tipped, 81
Secondary storage, 300–301, 302, 322
Selection function, 271
Self-diagnosis, 236
Serial, defined, 292
Sheep, 173–174
Shell, 120
Shell structure, 148–149
Shelter: agritechnology and, 157; need for, 12–13, 22, 27–28, 131–132
Shingles, defined, 144
Ships, 33, 37
Shopping by computer, 319, 330
Shortage(s), 55, 64; agritechnology and, 167–168; of farmland, 185
Skeleton, 212. *See also* Muscular-skeletal system
Skills: and adaptation, 311–312; changes in needs for, 308–310
Skyscrapers, 144–146; development of, 42; interior framework, 146
Slip rings, 125
Social Security, 263–264, 274
Socrates, 18
Socratic method, defined, 18
Software. *See* Computer *headings*
Soil analysis, 162
Solar energy, 50, 59, 113–114
Soldering, 83
Solution, technology as, 9
Source document, defined, 286, 295
Source transactions, 286, 295
South pole, 124
Soy beans, 161, 169
Space travel, 108, 338, 377; project structure, 347
Specialization, 26–27; in crop growing, 164–167, 180, 201
Species, defined, 161
Spinning, 30–31
Spinning wheels, 31
Spires, 135
Spray devices, 191–192
Sprinkler irrigation systems, 191–192, 204
Sprockets, 89
Square, defined, 140
Square structures, 140–141, 152
Starches, 240
Stationary, defined, 98
Stationary engines, 98
Steam engines, 36, 37, 45, 46, 110; development of, 32; principles of, 96–99; stationary, 98
Steel mills, location of, 33
Steering systems, 107
Stenographic training, 312–313
Sterilization of food, 176, 199–200
Stethoscope, 216
Stiffeners, 150
Stimuli, defined, 210
Storage (computer), 302; defined, 274; versus memory, 300–301; of records and files, 271–273. *See also* Main memory; Secondary storage
Storage of foods and goods, 160, 196–198, 204, 375
Straw, 196
Stress reduction, 249
Stretching, 244
Stringers, 150
Strokes (brain), 247–248
Strokes (steam engine), 97
Structural design, 41, 151, 152; of skyscrapers, 144
Structural forms, 134–142
Structure(s), 39–42; of buildings, 133–147, 151–152; defined, 41; and development of cutting tools, 78; exterior surface, 143, 144; form in relation to function in, 151; foundations, 133; free-standing, 134; multiple stories, 133, 143; perimeter-supported, 146, 152; permanent, 137–138; of vehicles, 148–149, 152; walls of, 143
Substance abuse, 225, 243
Sugar, 167–168, 180, 240
Sugar producing plants, 161, 167–168, 180
Suicide, 221
Sullivan, Louis, 42
Supermarkets, computers in, 259–261, 294, 295
Supplements, dietary, 242
Supply and demand, 55. *See also* Shortage(s)
Surgery, 229
Surplus, defined, 54
Surveying, laser beams used for, 185
Suspension: bridges, 147; system, 107
Sweet potatoes, 161
Synthetic fibers, 50
System(s): circulatory, 209; control functions in, 339–341; defined, 66; human body as, 214; information processing, 273; muscular-skeletal, 212; nervous, 210–211; respiratory, 210; of technologies, 66–67, 68, 95, 105–107, 222
System software, 284
Systems analysts, 278
Systems approach, 336, 338–345
Systems development, 336, 338–345

Tabulating machines, 264
Tap, defined, 82
Tape, magnetic, 291
Teamwork, 338, 351, 354, 357
Technical design, 359–360, 364
Technology: applications of, 10–21; and applied science, 336, 364; and change, 304–330; creation of, 335–365; defined, 3, 6, 21; development of, 25–27, 46, 222; effects of, 4–21; limitations on, 45–46; as major force, 45–46; progress and, 6–7; related systems of, 80, 105–107, 222; as a solution, 9; systems of, 66–67, 68, 95
Television, 16, 267, 383
Temperature: of combustion, 100–101, 104; of human body, 216, 232
Terraces/terracing, 185–186
Textile(s), 158
Textile mills, location of, 32, 33
Textile production, technology of, 31–32
Therapy: defined, 237; physical, 247–248
Thermometers, 217
Thread: cotton, 177; production of, 30–31
Three-dimensional diagnostic pictures, 230
Thresher, invention of, 196
Threshing, defined, 196
Thrust, 107, 108
Thrust engine, 107
Tides, as energy source, 62
Tilling, 203; purpose of, 186
Tomatoes, breeding and development, 202
Tools, 76–83; defined, 73, 90; examples, 74
Torque, defined, 89
Tracks: on magnetic disks, 292; on magnetic tape, 291
Tractor, 186
Trade-offs, 306–307, 352–353
Transaction(s), 273; defined, 268; records, 268–269; source, 286
Transcription, 312–313; of data, 286–287
Transferable skills, 311–313
Transmission of communication signals, 378

Transportation, 306, 373–375; agriculture and, 196–198, 204; air, 7–9, 37; defined, 33; of foods, 204; need for, 9, 10–12; systems, 66; technology of, 33, 36–38, 46; waterways, 33, 36. *See also* Aircraft; Railroads
Transposition of data, defined, 270
Trapezoidal structures, 142
Trauma, 221–222; defined, 221
Trauma care technology, 222
Treadle, defined, 31
Treatment, 230–231; defined, 230
Treatment, health care, 218
Tree crops, 161, 170–171, 180, 195
Trees: clearing of, 184, 203; pruning, 170
Triangular: roof forms, 136; shape, 134–136; structures, 137
Truck garden crops, 172–173
Trucking, 374
Turbine (jet) engines, 107–108, 111
Typesetting, 43, 315

Uniformity, 301; in information systems, 281–282
United States Bureau of Census, 45, 267, 274
Universal product code (UPC), 294
Unknowns, dealing with, 346, 348
Unskilled worker, defined, 309
UPC. *See* Universal product code
Urbanization, 27–28; and change, 306; defined, 25; structures and, 133–134; technological development and, 25–42, 46
Utilities, governmental control of, 55–56, 306
Utopia, defined, 18

Vaccines, 237, 250
VDT. *See* Video display terminal
Vegetables, 173
Vehicles: defined, 33; structure of, 148–149, 152
Veins, 209
Video display terminal (VDT), 298; eye strain and, 327
Video recording device, 318
Vinci, Leonardo da, 8, 19
Vine-type crops, 171–172
Viruses, 219
Vital (life) signs, 214–217, 230, 231, 237; diagnostic techniques, 227
Vitamins, 242, 250
Volatile, defined, 103

Wall covering, exterior, 143, 144
Wants, 15–16, 22
Warehouses, 375
Water, 242, 250; as energy source, 30, 31–32, 117
Water power, 30; looms driven by, 31–32
Water supply, 174, 178
Water table, defined, 191
Waterwheels, 30
Weapons, 10, 12, 77
Weather, 162
Weaving, 30
Weed control 188–190
Weight lifting, 247, 250
Welding, 83
Wheat, 163, 164; harvesting, 195–196; kernels, 196; milling, 198; planting, 188
Wheel: principle of, 85–86, 87; invention of, 306
Whitney, Eli, 178
Wind, as energy source, 30, 37
Windmills, 30, 59, 191
Wine, 172
Wood, 157, 178–179; as fuel, 62; as renewable resource, 64
Woodworking tools, 77–78
Word processing, 320
Work, 95–96; defined, 93, 110; plan of, 354–357
Work situation, ergonomic studies of, 327
Workers: displacement of, 308–311; unskilled, 309
World War II, information processing during, 265–266
Writing, 378–379

Xerographic copying device, 45
X-rays, 39, 229, 230, 235, 237; computer controls applied to, 230; sterilization of milk by, 176

Zone, defined, 56
Zoning laws, 40, 56
Zoning ordinances. *See* Zoning laws